T0347464

Routledge Revivals

Issues in U.S. International Forest Products Trade

After World War II, the United States became integrated into the world forest economy however the complexity of their trade agreements introduced several issues which needed to be addressed by world forestry policy. Originally published in 1981, this study delves into important issues related to forest resources and trade such as the future role of the United States in the world forest economy, trade restriction and U.S log exports. This title will be of interest to students of Environmental Studies and Economics.

Issues in U.S. International Forest Products Trade

Proceedings of a Workshop

Edited by
Roger A. Sedjo

First published in 1981
by Resources for the Future, Inc.

This edition first published in 2016 by Routledge
2 Park Square, Milton Park, Abingdon, Oxon, OX14 4RN
and by Routledge
711 Third Avenue, New York, NY 10017

Routledge is an imprint of the Taylor & Francis Group, an informa business

© 1981 Resources for the Future, Inc.

All rights reserved. No part of this book may be reprinted or reproduced or utilised in any form or by any electronic, mechanical, or other means, now known or hereafter invented, including photocopying and recording, or in any information storage or retrieval system, without permission in writing from the publishers.

Publisher's Note
The publisher has gone to great lengths to ensure the quality of this reprint but points out that some imperfections in the original copies may be apparent.

Disclaimer
The publisher has made every effort to trace copyright holders and welcomes correspondence from those they have been unable to contact.

A Library of Congress record exists under LC control number: 80008885

ISBN 13: 978-1-138-95265-2 (hbk)
ISBN 13: 978-1-315-66758-4 (ebk)

Issues in U.S.
International Forest Products Trade

Research Paper R-23

Issues in U.S. International Forest Products Trade

Proceedings of a Workshop Held in Washington, D.C.
on March 6 and 7, 1980

Roger A. Sedjo, editor

RESOURCES FOR THE FUTURE / WASHINGTON, D.C.

RESOURCES FOR THE FUTURE, INC.
1755 Massachusetts Avenue, N.W., Washington, D.C. 20036

DIRECTORS: M. Gordon Wolman, *Chairman,* •Charles E. Bishop, Roberto de O. Campos, Anne P. Carter, Emery N. Castle, William T. Creson, Jerry D. Geist, David S. R. Leighton, Franklin A. Lindsay, George C. McGhee, Vincent E. McKelvey, Richard W. Manderbach, Laurence I. Moss, Mrs. Oscar M. Ruebhausen, Janez Stanovnik, Charles B. Stauffacher, Carl H. Stoltenberg, Russell E. Train, Robert M. White, Franklin H. Williams. *Honorary Directors:* Horace M. Albright, Erwin D. Canham, Edward J. Cleary, Hugh L. Keenleyside, Edward S. Mason, William S. Paley, John W Vanderwilt.

OFFICERS: Emery N. Castle, *President;* Edward F. Hand, *Secretary-Treasurer.*

Resources for the Future is a nonprofit organization for research and education in the development, conservation, and use of natural resources and the improvement of the quality of the environment. It was established in 1952 with the cooperation of the Ford Foundation. Grants for research are accepted from government and private sources only if they meet the conditions of a policy established by the Board of Directors of Resources for the Future. The policy states that RFF shall be solely responsible for the conduct of the research and free to make the research results available to the public. Part of the work of Resources for the Future is carried out by its resident staff; part is supported by grants to universities and other nonprofit organizations. Unless otherwise stated, interpretations and conclusions in RFF publications are those of the authors; the organization takes responsibility for the selection of significant subjects for study, the competence of the researchers, and their freedom of inquiry.

Research Papers are studies and conference reports published by Resources for the Future from the authors' typescripts. The accuracy of the material is the responsibility of the authors and the material is not given the usual editorial review by RFF. The Research Paper series is intended to provide inexpensive and prompt distribution of research that is likely to have a shorter shelf life or to reach a smaller audience than RFF books.

Library of Congress Catalog Card Number 80-8885

ISBN 0-8018-2634-9

Copyright © 1981 by Resources for the Future, Inc.

Distributed by The Johns Hopkins University Press,
  Baltimore, Maryland 21218

Manufactured in the United States of America

Published February 1981. $10.50.

TABLE OF CONTENTS

ACKNOWLEDGMENTS

A variety of individuals and organizations cooperated in the successful presentation of this workshop and the publication of these proceedings. First and foremost was the co-sponsor of the workshop, the American Forestry Association (AFA). A special thanks goes to the staff of the AFA and particularly to Rex Ressler and Dick Pardo.

Of course, the workshop could not have been possible without the contributions of the various papergivers and reviewers. A listing of the contributors appears in the Appendix. In this regard, a special acknowledgment is in order for Emery Castle, Rex Ressler, and Marion Clawson for their important contributions as session moderators.

In addition, an appreciation must be extended to the Weyerhaeuser Company Foundation and the U.S. Forest Service for their support of the work of the Forest Economics and Policy Program at Resources for the Future and this workshop.

Finally, a very special thanks to Ken Frederick for his review of the workshop papers, to Avery Gordon for her assistance in the organization of the workshop and typing of this report, and to Marian Lesko for her typing efforts.

R.A.S.

1

INTRODUCTION AND SUMMARY

On March 6 and 7 of this year Resources for the Future (RFF) and the American Forestry Association (AFA) jointly sponsored a workshop on Issues in U.S. International Forest Products Trade. The workshop was the out-growth of the organizations' common concern that insufficient attention was being given to this important topic. And, indeed, until recently this was so.

Historically, the United States has viewed itself as being basically self-sufficient in forest products. It is true that as long ago as World War I the United States had become a net importer of lumber, but the trade flows that developed then were largely with Canada, and the broader global questions and issues had not surfaced.

However, this narrow perspective increasingly became inappropriate as the United States became more extensively integrated into the world forestry economy during the period following World War II. In the 1950s, U.S. imports of tropical hardwood products began to grow, and the early 1960s saw the development of a large external market for U.S. conifer sawlogs in Japan. Also during this period U.S. pulp and paper exports became significant earners of foreign exchange although the United States continued to rely heavily on foreign suppliers for newsprint, lumber, and other forest products. Indeed, the United States continued to experience net trade deficits in forest products that often were in excess of $1 billion.

The rising volume and complexity of these trade flows could not help but generate a variety of new policy issues relating to U.S. forest products. In response, research into these issues has increased considerably in recent years, but even today very few organizations and individuals within the United States are studying these questions. One agency, the U.S. Forest Service, has undertaken systematic research in this area, and in recent years the service has published numerous studies, many of them originating from its Pacific Northwest Experiment Station. The recently developed Forest Service Timber Assessment Model has been utilized to investigate international forest trade questions. Also, within the past three years RFF has undertaken research in international forestry issues as part of its expanded Forest Economics and Policy Program.

The purpose of the RFF/AFA workshop was to provide an opportunity for researchers and interested individuals to focus upon a particular set of issues in international forest products trade. We chose issues that are timely in a policy sense, and also ones that have recently experienced intensive research. Without exception, the papers and research results presented at the workshop are the product of current research efforts.

The fraternity of researchers examining international forest products trade is still small, which made it relatively easy to involve most of them in the workshop. In addition to those doing research, those attending the workshop included industry representatives, Forest Service employees, and other interested individuals from the private sector and academe. It was hoped that during the workshop the interaction among this group would generate a dialogue that would provide additional insights into the nature of these issues and contribute to the quality of the public policy debate.

The technical level of the papers varied greatly, but generally the papers became more technical as the workshop progressed into more specialized areas. The papers included in this volume are primarily directed at individuals interested in forestry trade issues. It is assumed that the reader will have some familiarity with both forest products and economics.

Although the focus of the workshop was on forest resources and the forest products trade, many of the general issues faced within this sector are similar to those confronted by other resources. For example, restricting the export of a primary product in an attempt to reduce costs for domestic processes is an issue faced not only by U.S. log producers but also by the U.S. scrap steel industry. Other similarities abound. Thus, although the principal audience for these papers is likely to be drawn from the forestry sector, economists, resource managers, and others interested in resource trade are likely to find that the types of issues examined and the technical methodologies employed here may have substantial applicability to other resource areas as well.

Obviously, not every international forestry issue of current interest to the United States could be addressed in one brief workshop. Of necessity, the focus was narrowed, and three related topics were addressed. The first session was the broadest, examining what is perhaps the most nebulous, but also the most important, long-run issue; the future role of the United States in the world forest economy. The second session examined the effect on U.S. trade of changing forest products trade restrictions, especially in light of the recent agreement arising from the Multilateral Trade Negotiations (MTN). The final session, which also examined the effect of trade restrictions focused exclusively upon the narrow but important forest trade policy questions related to U.S. log exports and the economic implications of various restrictions on such exports.

Each session began with the presentation of two or three papers which were then critically reviewed. The presentation and reviews provided a backdrop for the subsequent discussion by selected panelists and the audience at large. The approach proved to be effective, and each session was characterized by wide-ranging discussions that followed the somewhat more narrowly focused formal presentations.

The first session examined the future U.S. role in the world forest resource trade. Many have claimed that the forests and forestlands of the United States have the potential to become the world's "woodbasket." Despite these claims, the United States has been, and continues to be, a net importer of forest products. Why is there a discrepancy between its perceived potential and realized performance? While this question is obviously too complex and multidimensional to be wholly examined in a single session, the answer undoubtedly involves a greater knowledge of the intricate global interrelationships of production and trade.

Two of the papers examine some of the facets of this problem. Roger Sedjo's paper hypothesizes that the world is currently experiencing a transition from natural to plantation forestry, not unlike that which took place earlier when humans moved from gathering and hunting to agriculture and livestock raising. As a case study, Sedjo estimates quantitatively the production and export potential of Brazil, the single country generally believed to have the greatest potential impact upon the world market for a specific forest product. Sedjo's results suggest that curren

nd future Brazilian forest plantations have the potential to generate
ignificant impacts upon the structure of world chemical pulp production
nd trade flows as early as the mid-1990s. More generally, the quantifi-
ations of the paper are indicative of the potential wide-ranging impact
hat these exotic plantations may have on world markets.

David Darr's paper is an outgrowth of the Forest Service effort to
evelop a comprehensive Timber Assessment Model. Darr discusses the
orest Service's long-term projections and the methodology used in ob-
aining those projections. It is interesting to note that this methodology
oes not employ sophisticated forecasting techniques but rather involves a
ystematic extrapolation of existing trends while making ad hoc adjust-
ents for such factors as population, housing starts, and anticipated
hanges in output from both domestic and foreign sources. In conclusion,
arr states that generally, "Our projections indicate a continuation of
urrent U.S. trade patterns for timber products."

These papers were then commented on by Jan Laarman, John Zivnuska,
ohn Ward, and Dwight Hair (whose comments are not included in this
olume). While Laarman's comments generally support Sedjo's view that
here exists a great potential for timber plantations in particular
ountries to affect future wood supplies, he demonstrates that only a
ery few countries have the potential to be major forest products ex-
orts by utilizing newly created plantations. Laarman also stresses the
ariety of domestic uses that might claim the timber and notes the diffi-
lties involved in converting standing timber into an internationally
adable commodity.

John Zivnuska, commenting on Darr's paper, notes an inconsistency between the price assumptions of the trade projections and the price assumptions of the Timber Assessment Model. The trade projections assume that the base period prices prevailed continuously, whereas the Timber Assessment Model predicts rising prices. Zivnuska points out that the failure to utilize the predicted higher prices in the trade projections creates a systematic bias toward projecting a larger trade deficit than is consistent with the prices projected by the Timber Assessment Model.

John Ward's comments are also directed toward Darr's paper and the trade projections used in the Timber Assessment Model. Ward argues that the projections of U.S. forest product exports are too conservative since they (1) tend to underestimate future U.S. supplies; (2) do not adequately consider growth in overseas markets; and (3) do not fully appreciate the inducement to exports created by the overseas profit potential.

To summarize the session's findings: regardless of the Forest Service projections, the issue of the magnitude of U.S. forest product net trade continues unresolved. Zivnuska makes it clear that the Forest Service projections systematically overstate the probable U.S. forest products trade deficit. Conceivably, the bias could be so large that more careful projections using equilibrium rather than the base period might predict an overall forest product surplus. As Zivnuska notes, fifty-year projections are very tenuous, and other factors will dominate beyond the first twenty years. Sedjo's projections for Brazil suggest the possibility of one such other factor, that is, major alternative world suppliers

The afternoon session addressed the issue of restrictions on international forest products trade and focused upon changing trade restrictions, particularly those arising out of the Multilateral Trade

Negotiations (MTN). Two papers dealing with the effect of trade restrictions upon U.S. forest products trade were presented. The first, by Darius Adams and Richard Haynes, compares the differential effects of the separate imposition of a hypothetical tariff and quota upon the importation of lumber into the United States. The second paper by Samuel Radcliffe discusses the recent MTN agreement and its likely effect upon U.S. exports and imports in forest products.

The Adams-Haynes paper utilizes the Forest Service Timber Assessment Model to simulate the effect of the restrictions upon the various U.S. regions and to present projections of regional lumber and stumpage prices, trade volumes, income distribution, and economic efficiency. The paper examines the market and resource impacts resulting from the hypothetical imposition of an ad valorem tariff on lumber imports from Canada and the effects of a fixed, annual lumber-import quota of 11 billion board-feet. Adams and Haynes concluded that restrictions of both types will lead to higher rates of stumpage price growth, generating benefits to producers, higher consumer prices, and higher levels of domestic harvest. They also conclude that a lumber import restriction, of either type, would result in "reductions in consumption (that) would be concentrated in the eastern U.S. while expanded domestic production would come primarily from the South."

Samuel Radcliffe's paper details the changes in restrictions on forest products resulting from the MTN agreement and gives a quantitative estimate of the changes in trade volumes that are likely to occur. Radcliffe concludes that the aggregate effect upon production and trade in U.S. forest products is likely to be small, although for individual firms and specialized commodities it could have an important effect.

These papers were followed by comments from A. Clark Wiseman, Louis Vargha, Harold Wisdom, and Irene Meister (whose comments are not included in this volume). While not challenging the basic nature of the interregional adjustment projected in the Adams and Haynes paper, Wiseman warns against confusing the consistency of a model's results with its accuracy as a predictive tool and suggests that many of the predictions are built into the structure of the model. He also challenges the quantitative welfare estimates of the model since they indicate that the effect of this trade restriction is to increase U.S. welfare, whereas it is generally agreed that the imposition of a trade restriction, in the absence of very special conditions, will create net welfare losses. Wiseman suggests that the difficulty may lie with the improper specification of short- and long-run curves.

In his comments on Radcliffe's paper, Vargha expresses his belief that the United States has not gained much from the MTN agreement and thus is in basic agreement with Radcliffe. However, Vargha notes that Radcliffe's approach, which was designed to focus solely upon the effect of the changes in trade restrictions, is therefore unable to assess the various other dynamic real-world forces that are likely to be present. Vargha also indicates that the MTN agreement signaled to the various producers in participating countries just which industries could expect future protection and which could not.

Although Wisdom also basically agrees with Radcliffe, he believes that the effects of the MTN would differ somewhat from what Radcliffe suggests. He attributes some of this difference to the fact that the two-country model used by Radcliffe is an imperfect tool with which to address the complexities of multilateral tariff reductions. In general, however, Radcliffe's aggregate quantitative estimates were viewed by Wisdom as reasonable ones.

To summarize the conclusions of this session, there was a consensus that the effect of the MTN upon U.S. forest products trade would be small and that the predictions of Radcliffe's approach were, by and large, reasonable; especially in the aggregate. The modest size of the effects reflects the fact that the negotiated restriction reductions were small and many unquantitative restrictions remained in place. As is usually the case with trade restrictions, the effects of the simulated lumber trade restrictions between the U.S. and Canada were to reduce the trade flow, to raise the price in the importing country while reducing the price in the exporting country, to cause a domestic restructuring of consumption and production, and to generate redistribution and net welfare effects. The participants' primary disagreement was with the atypical gains in net welfare attributed to the tariff. Wiseman points out (and correctly so, I believe) that this outcome is the result of the model's applying essentially short-run supply curves to a problem that requires long-run supply curves.

The final session of the workshop was devoted to examining the effect of restrictions on a particular commodity. The log export issue has been in the forefront of the policy debate for almost two decades. Simply stated, the issue is the extent, if any, to which the United States should further restrict its conifer log exports from the Pacific Northwest to East Asia, and the economic effects of additional restrictions, particularly an export prohibition. Log exports began in earnest in the early 1960s after the great Columbus Day storm damaged the forests of the Pacific Northwest region. The resulting salvage operation stimulated the then fledging log exports to Japan. A few years later, in 1968, the Morse Amendment was enacted, restricting log exports from federal lands. This restriction was further tightened in 1974.

The issue is complex and multidimensional. What effect have current restrictions upon domestic production of logs, lumber, and other forest products? What would be the effect of further restrictions? To what extent would the Japanese purchases of foreign lumber increase in response to log export prohibition and what impact might this have upon domestic U.S. production and prices of both logs and lumber? Who would benefit and who would experience losses from such a prohibition? What would be the effect on the net income of the Pacific Coast region and on other U.S. regions?

Two papers presented at this session applied models to determine the economic impact of a total export ban, a restriction that often has been advocated, as compared with a situation of unrestricted trade. A. Clark Wiseman and Roger Sedjo discuss the commonly accepted concepts of Marshallian consumer and producer surpluses and develop a model utilizing those concepts. The model examines the income distribution and economic efficiency implications of a ban upon log exports from the Pacific Coast region. The paper by Richard Haynes, David Darr, and Darius Adams utilizes the Forest Service Timber Assessment Model to examine the implications of a log export ban in a multiregional context. Although the two models are quite different in design and structure, their projections are

trikingly similar. Both models recognize that the tight interrelation-
hips between the log and lumber (processed wood) markets implies that
anning log exports would result in a shift of foreign demand from logs
o lumber, thereby moderating or wholly negating both the anticipated in-
rease in domestic lumber availability and the expected declines resulting
rom a log export ban. The maximum long-term lumber price decrease pro-
ected by both models was 3.7 percent, and the Wiseman-Sedjo model even
uggests the possibility of a very slight long-term lumber price increase.
he Haynes-Darr-Adams model projects a short-term (several year) lumber
rice _increase_ of as much as 9.87 percent because of the short-term limita-
ions of processing capacity. Both models also project decreases in
acific Coast stumpage prices and increases in lumber processing.

Finally, both models project net losses in economic welfare, that is,
egional or national income, and large wealth transfers primarily from
esources used in log production to resources used in log processing.
he Wiseman-Sedjo model predicts annual welfare losses of up to $50 million
hile the model developed by Haynes, Darr, and Adams estimates the annual
osses (by 1990) to be at least $141.7 million. The Wiseman-Sedjo model
stimates a wealth transfer within the Pacific Coast region of up to
?02.8 million annually, whereas the Haynes-Darr-Adams model projects the
ationwide transfer to be over $500 million annually by 1990. In addition,
iseman and Sedjo note that whereas the ban would probably result in
igher employment in the processing industry, increases in industry em-
loyment should not be confused with overall increases in regional em-
loyment.

In a third paper, Barney Dowdle addresses the question of log export

restrictions within the broader context of the total supply potential
of the Pacific Northwest and the ceilings imposed by the nondeclining
even-flow constraints which limit the rate of which old growth is harvested in the West. Dowdle argues that log exports are a problem only
because the limits imposed on the federal forest harvest levels artifi-
cially constrain the domestic log supply. Dowdle also states that the
current prohibition on the export of logs from federal forests may have
caused a situation in which higher prices can be charged for exported log
thereby allowing U.S. exporters to collect economic rents from East Asian
log buyers, in a manner akin to the charging of higher prices by the
oil-exporting nations.

Philip Cartwright, in his discussion of the Wiseman-Sedjo paper,
suggests some difficulties with the model, especially its assumption of
homogeneity of outputs, particularly of lumber. Cartwright points out
that from the extent to which domestic and foreign lumber are not readily
substitutable, the welfare loss that such a log export ban imposes on the
Pacific Coast is likely to be greater than that estimated by Wiseman and
Sedjo. In addition, Cartwright notes an additional economic loss to consumers that is not discussed in the paper. To the extent that a ban woul
depreciate the U.S. dollar as a result of decreased net U.S. foreign exchange earnings, Americans consuming imports of any type would experience
losses of consumer welfare via the resulting higher dollar prices necessary for the purchase of imported items.

In Bruce Lippke's review of the Haynes-Darr-Adams paper, he states
that although the model represents an improvement in modeling technique,
it has been designed primarily to examine domestic markets and

interactions and, therefore, is less well suited to dealing with international issues. Lippke argues that the authors have not paid sufficient attention to the massive wealth transfers that would accompany a ban. A major point made by Lippke is that a log export ban, which would reduce stumpage prices and result in large wealth transfers away from stumpage owners, would thereby reduce economic incentives for intensive forest management and ultimately reduce total wood availability, total forest industry investment and employment.

William McKillop, in the final review of the workshop, raises several questions regarding Dowdle's paper. Acknowledging Dowdle's contention that a higher federal timber harvest would lessen the pressure for an export ban, McKillop notes, "In dealing with the log export questions, per se, it is almost essential to take National Forest output policy as given if issues are to be analyzed in a meaningful way." McKillop also offers an alternative interpretation of the causes of the log price differential between domestic and exported logs, stressing differences in quality and transport costs as explanatory factors accounting for the observed price differentials.

In summary, the papers reach a high degree of consensus in their assessment of the effect of a log export prohibition. Where comparable, the price, quantity, redistribution, and welfare effects were generally in the same direction and of roughly similar magnitudes, for example, relatively large or relatively small. Dowdle's analysis, which is different in approach and focus, raises the possibility that existing restrictions result in the creation of markets with some noncompetitive elements which in turn may have generated higher prices for exported logs.

The workshop ended, leaving us with new insights into the various problems and issues that had been raised. Our trust is that our more sophisticated understanding of the nature of these complex issues will contribute to better policy decisions in the future. It is to this end that the trade workshop was undertaken.

Washington, D.C.
November 1980

Roger A. Sedjo, Director
Forest Economics and Policy Program
Resources for the Future

PART I

THE FUTURE U.S. ROLE IN WORLD FOREST RESOURCE TRADE

WORLD FOREST PLANTATIONS--
WHAT ARE THE IMPLICATIONS FOR U.S. FOREST PRODUCTS TRADE?

Roger Sedjo

The theme of this session is The Future U.S. Role in World Forest Re-
ource Trade.   Obviously, there are a variety of factors that can affect
he future trading role of the United States.   One development that is not
et well recognized as having the near term potential to importantly affect
.S. trade in forest products, and indeed to influence the worldwide struc-
ure of forest products trade, is the creation of exotic forest plantations
wly located in regions that have not traditionally been important produ-
rs and exporters.

This paper will briefly examine the potential of these plantations to
ffect future world markets for industrial wood products, First, the nature
  the transition to plantations will be examined, the major types of plan-
tion activities that are now occuring will be characterized, and recent
rest plantations' experience in a number of important world regions will
  summarized. Next, the paper will examine, as a case study, the exper-
nce of Brazil in establishing rather massive land areas in exotic forest
antations. Third, a crude estimate of the possible impact upon a major
rld market of the Brazilian wood volumes that could be forthcoming by the
d 1990s will be made. Finally, the implications of plantation activities

from nontraditional wood-exporting countries upon U.S. international trade
forest investments, and forestland use will be briefly discussed.

Background

Although plantation forests make up only a small fraction of the
world's total industrial wood supply, they are becoming increasingly impor
tant as the natural forests of the world decline, reflecting utilization o
the natural stands or land clearing for nonforest uses. As this process
continues, greater portions of the remaining commercial forests represent
regrowth, much of it as the result of the conscious intervention of man.
Commercial plantation forests, that is, those planned and actively managed
for their commercial wood values, are gradually replacing the natural forests in many regions of the world, and exotic species, that is, non-indige
eous species, are being introduced anew in other regions.

Historically, human needs for wood have been met in a manner akin to
the hunting and gathering mode employed by early man to meet his food need
Just as hunting and gathering have been almost entirely replaced by agricu
ture and livestock raising, world wood requirements are increasingly being
met from managed plantation forests rather than unmanaged natural forests.
The transition from natural to plantation forests has, thus far, created
only small changes in world trade patterns since the principal plantation
developments have been in regions where plantations have merely replaced
natural forests. However, considerable forest plantation activity is now
under way in many regions of the globe. Some of this activity is likely
to have important impacts on future world trade patterns.

Major plantation activities can be characterized as occuring in three
different regional settings. First, in temperate regions that traditional

roduced the majority of the world's industrial wood--northern Europe and
North America--plantations utilizing indigeneous species have typically replaced cutover natural forests. Second, other temperate regions that have
not been traditionally major industrial wood producers are commonly utilizing in their plantations exotic temperate climate species (largely North
American) that exhibit rapid growth and desired merchantibility. Third,
certain tropical regions are introducing exotic species (tropical pines,
eucalyptus, gmelina) from other tropical regions which exhibit the desirable
growth and merchantibility characteristics. While experience with exotic
plantations in the tropics is limited, results thus far are so dramatic
that some knowledgeable observers maintain that tropical regions will even-
ually become dominant wood suppliers.[1]

It should be noted that although plantations are a small fraction of
the world's forested area, the current land areas involved belie their true
potential. Industrial potential is the result of not only more land being
converted into forest plantations, but also of the volumes of output per
land unit. These volumes are likely to be large for plantations since the
location is usually determined, at least partially, by considerations of
high biological growth and also because management practices usually associated with plantations increase usable growth.

---

[1] See Johnson (1976) for a discussion of promise and problems of tropical plantations. For an optimistic view of the potential for tropical plan-
ations, see Kellison (1979).

Plantation Experience

Plantation forests of some type exist in almost every country. However, commercial plantations are important in only a relatively few regions.
In Europe and North America--the world's major supply source of industrial
wood--the natural forests are gradually giving way to managed plantations.
The process is most advanced in Europe, except for the USSR, where plantation forests have been maintained for hundreds of years. For example, the
United Kingdom is currently aforesting some 30 thousand hectares per annum
and anticipates large output increases in the twenty-first century. In
North America, plantation forests developed primarily in the U.S. South in
the 1930s and have become increasingly important since World War II. Initially, U.S. plantations involved merely reforestation with little, if any,
subsequent management. However, in recent years in the U.S. South and Pacific Northwest, commercial plantations utilizing a range of practices have
become prevalent. By contrast, plantation forestry in Canada is still in
its fledging stages caused in part by the large inventories of natural
forest still available.

Outside of Europe and North America, several regions have developed
large commercial plantation forests. New Zealand currently has about one
million hectares in plantation forests, most of it in exotic conifer from
North America. New Zealand is adding to these forests at a rate of some
50,000 hectares per year. Australia has also introduced exotic plantations
particularly conifers, on a large scale to provide for her domestic long
fiber requirement. Plantation forestry of various types is under way in
Japan, the Philippines, Indonesia, and Malaysia. India has undertaken
large-scale plantation forestry activities. Turkey and Iran have also

undertaken significant plantation activities with a view to commercial industrial outputs. Numerous parts of Africa have been involved in various
forestry schemes. Plantations have been created in Kenya and Tanzania in
East Africa, utlizing a variety of tropical conifer species. South Africa
has undertaken large plantations of various species, especially P. patula.
West Africa has been involved in plantation schemes utilizing eucalyptus,
gmelina, and tropical pines.

Of all the nontraditional producing areas of the globe, however, commercial forest plantations have become most important in South America.
Plantations of P. radiata have been in place in Chile for about seventy-
five years. In recent years the planting activity in Chile has increased
substantially, with a governmental goal of 80,000 hectares per year, and
within the past five years Chile has begun to actively enter world markets.
Significant plantation activity is also under way in Colombia and Argentina.
In Venezuela, serious interest in large commercial plantations has developed
only within the past decade or less. However, the level of activity is substantial involving hundreds of thousands of hectares and appears to have
great potential due to the availability of large areas of land with few alternative uses and a location astride a major navigatable artery which provides access to the sea.

Brazilian Experience: A Case Study

Nowhere in the world, however, has exotic plantation forestry been so
active in recent years, nor has it shown more potential than in Brazil
(Zobel, 1979). The potential of Brazil is vast for several reasons. First,
biological growth rates in Brazil are very rapid compared with those experienced in North America (Zobel, 1979, p. 5). Second, the land areas of

Brazil that are available for forest plantations are vast, and in many cases the opportunity costs of the land are quite small. Third, as will be discussed, the government of Brazil has deemed it appropriate to heavily subsidize the creation of new plantations. Whatever the underlying economics of the plantation investments may be, once the plantations are in place the economics of harvesting and processing are likely to be generally favorable. Finally, the favorable biology implies a short rotation, about seven years for eucalyptus and twelve years for pine. Thus, current plantings ca be harvested in the late 1980s or early 1990s.

Brazil first introduced eucalyptus from Australia in the early twentieth century as fuel for the railroads and hence located plantations along the railways lines. Subsequently, eucalyptus was found to lend itself to the production of charcoal, which could be substituted for coking coal in the production of steel. With large deposits of iron ore and limited cokin coal, charcoal produced from plantation eucalyptus has become important to the Brazilian steel industry.[2] More recently, eucalyptus has been utilized quite successfully as the pulpwood for wood pulp production. In the 1950s, plantations of U.S. southern pines, principally slash and loblolly, were introduced into southern Brazil, and flourished. However, it was with the Brazilian fiscal incentives program, beginning in the mid-1960s, that massive establishment of plantation forests began.

Incentives Program

The fiscal incentives program allow industrial companies to substitute investments in governmentally approved investment areas for the payment of

---

[2]Charcoal is also required to make certain types of high-quality steel

to 50 percent of their tax liability (Beattie, 1975). Forest plantation
establishment is among the approved investments. In essence the corporation
can either pay its taxes, receiving nothing tangible in return, or can invest those funds in an approved investment, thereby accumulating an asset.
Tables 1, 2, and 3 show the amount of forest plantation establishment, in
pine and eucalyptus, under the incentives program. In the twelve years en-
compassed between 1967 and 1978, a total of about 3 million hectares of land
were put into plantation. This combined with an earlier plantation area of
about 0.5 million hectares to give a total of about 3.5 million hectares in
1978. Furthermore, preliminary estimates suggest that reforestation for
1979 approached 0.5 million hectares. Also impressive is the fact that the
total area being reforested and aforested showed substantial increases over
the period even as the fiscal incentives were reduced somewhat. As table 1
indicates, eucalyptus constitutes over 60 percent of the newly created plantations, with pine being about 35 percent and small volumes of native parana
pine (Araucaria) and other species making up the rest. It should be noted
that these figures do not include about 100,000 hectares of large plantation activities undertaken in the Amazon that are not under the fiscal incentives program.

As substantial as plantation efforts have been in the past decade and
half, the expectations for the near future promise even greater planting
levels. The first six years of the 1980s are expected to see the creation
of about 0.5 million hectares of new forest plantations each year (Ministerio da Agricultural, 1978).

Table 1. Brazilian Plantation Forests, 1967-78

(thousand hectares)

State	Eucalyptus	Pine	Other	Total
Bahia	24.872	57.027	- [a]	81.899
Distrito Federal	11.271	4.103	-	15.374
Espírito Santo	128.551	1.214	.045	129.810
Goiás	31.803	8.348	1.400	41.551
Mato Grosso do Sul	306.128	31.437	-	337.565
Maranhão	.010	-	-	.010
Minas Gerais	817.525	141.296	.560	959.381
Paraná	48.834	346.815	45.251	440.900
Rio Grande do Sul	18.268	89.216	7.912	115.396
Rio de Janeiro	9.880	1.626	.015	11.521
Santa Catarina	13.381	231.970	14.688	260.039
São Paulo	329.035	191.986	3.204	525.225
Total	1.739.558	1.105.038	73.075	2.917.671

Note: These are plantations eligible for fiscal incentives program.

Source: Relatorio Estatistico--1978, (São Paulo, Brazil, Associacao Natural dos Fabricanotes de papel e Celulose).

[a]Dashes denote zero value.

Table 2. Eucalyptus Plantation Establishment

(thousand hectares)

Eucalyptus	1967	1968	1969	1970	1971	1972	1973	1974	1975	1976	1977	1978	Total
Bahia	-	-	-	.050	-	2.200	4.802	.562	.620	3.961	4.213	8.464	24.872
Distrito Federal	-	-	-	-	-	-	-	-	-	2.829	4.720	3.722	11.271
Espírito Santo	.020	.832	2.761	8.376	9.135	27.141	23.175	23.407	11.548	17.706	2.473	1.977	128.551
Goiás	.038	.898	1.239	20.096	3.135	2.614	2.902	4.722	3.290	5.833	2.074	2.962	31.303
Mato Grosso do Sul	-	-	-	.894	6.243	10.901	19.713	35.556	69.669	79.615	40.986	42.641	306.128
Maranhão	-	-	-	.010	-	-	-	-	-	-	-	-	.010
Minas Gerais	8.364	12.164	22.552	35.884	41.663	52.708	56.280	80.210	100.573	128.141	124.141	154.137	817.525
Paraná	1.051	.783	2.285	1.459	8.156	11.533	2.961	6.225	4.161	5.423	2.774	2.022	48.834
Rio Grande do Sul	.004	.068	.255	2.270	3.270	5.019	1.543	1.194	.785	2.180	.009	1.671	18.268
Rio de Janeiro	.400	.235	.915	1.307	1.073	.921	1.449	.422	2.406	.252	-	-	9.380
Santa Catarina	.066	.099	.363	.554	1.671	.758	1.478	2.698	2.990	1.886	.647	.171	13.381
São Paulo	3.933	14.478	23.432	30.710	54.706	58.646	46.880	33.340	26.675	14.510	11.445	10.180	329.035
Total	13.876	30.057	53.802	83.610	129.052	172.441	161.183	188.337	222.717	262.336	194.100	228.047	1.739.553

Note: Dashes denote zero value.

Source: Relatorio Estatistico—1978, (São Paulo, Brazil, Associacao Natural dos Fabricanotes de Papel e Celulose).

25

Table 3. Pine Plantation Establishment

(thousand hectares)

Pine	1967	1968	1969	1970	1971	1972	1973	1974	1975	1976	1977	1978	Total
Bahia	-	-	-	.003	.843	-	1.818	2.962	11.963	13.010	9.678	16.750	57.027
Distrito Federal	-	-	-	-	-	-	-	-	-	-	.577	3.526	4.103
Espírito Santo	-	.027	.086	.324	.427	.208	.120	-	.022	-	-	-	1.214
Goiás	-	-	-	-	-	.018	.075	.005	-	-	-	8.250	8.348
Mato Grosso do Sul	-	-	-	.785	.980	-	.824	.905	1.749	6.980	7.605	11.619	31.437
Minas Gerais	.797	.821	.713	7.052	7.582	7.087	14.773	10.036	11.614	22.051	24.517	35.533	141.296
Paraná	4.913	16.658	35.696	44.118	30.443	31.063	26.431	26.431	31.588	34.737	24.183	39.077	346.815
Rio Grande do Sul	.145	3.038	7.215	7.123	8.721	17.195	5.589	3.304	5.505	7.710	10.339	8.332	89.216
Rio de Janeiro	.185	.150	.223	.130	.228	.062	.166	.402	.080	-	-	-	1.626
Santa Catarina	5.111	15.285	19.637	21.245	22.699	23.094	22.630	20.613	25.278	16.786	21.520	18.072	231.970
São Paulo	7.008	23.921	33.228	39.135	26.139	21.612	13.754	13.110	6.421	5.728	.364	1.566	191.986
Total	18.159	60.900	96.798	119.915	98.052	101.059	86.180	83.245	94.220	107.002	98.783	140.725	1.105.038

Note: Dashes denote zero value.

Source: Relatorio Estatistico—1978, (São Paulo, Brazil, Associacao Natural dos Fabricanotes de Papel e Celulose).

26

Distribution of Plantation Activities and Growth Performance in Brazil

By and large the aforestation activities in pine have been undertaken in the south of Brazil in the states of Paraná and Santa Catarina at latitudes roughly equivalent to that of Florida. The most commonly utilized species in this region is slash (P. elliottii), and loblolly pine (P. taeda), with loblolly predominating. For an average site the expected growth rate is about 20 cubic meters of solid wood per hectare[3] per year.[4] To date, activities have largely been directed to wood production for pulp and paper mills; however, there is increasing interest in the production of various types of solid woods including lumber and veneers.

As one moves north across the Tropic of Capricorn, the pine plantations give way to eucalyptus. An important constraint toward the southern edge of the eucalpytus region is the difficulty encountered with frost damage to eucalyptus. Efforts are under way, however, to find frost-resistant eucalyptus, and some progress has been made. This region--consisting of the states of São Paulo, Minas Gerais, Espírito Santo, and Mato Grosso do Sul--is larger than the Old South of the United States at a latitude roughly comparable to that of Cuba and the Yucatan peninsula of Mexico.

Here, eucalyptus plantings (including importantly E. grandis, E. urophylla and E. saligna) far exceed the plantings of pine. The pine that is planted in this region is predominantly tropical, such as P. caribaea, P.

---

[3]Twenty cubic meters per hectare $\approx$ 267 cubic feet per acre.

[4]The growth rates here are relatively conservative and represent what appear to be "typical" plantation yields as opposed to experimental or best site performances. For example, much higher yields are reported by the Instituto de Pesquisas e Estudo Florestais (IPEF, n.d.). In addition, genetically improved trees offer promise of substantially higher yields within the next decade.

kesiya and P. oocarpa. By far the dominant state experiencing aforestation is Minas Gerais, which is also Brazil's steel-making center. Almost one-third of the total new plantations created between 1967 and 1978 occurred in this state. The location of forest plantations in the steel-making region is not simply coincidence, since it is estimated that about 80 percent of the wood produced from plantations in Minas Gerais is used for the production of charcoal which is used in steel-making. In contrast, the dominant use of plantation wood in the other states throughout Brazil has been for traditional commercial purposes, and particularly for pulp production. Growth rates of eucalyptus have been reported in one state in this region as in excess of 35 cubic meters per hectare per year. An overall growth rate for euclayptus of 25 cubic meters per hectare per year appears to be a conservative average (Carbonnier and Lonner, 1976).

Moving further north, the level of plantation activity declines dramatically. Along the Atlantic coast in Bahia State, recent changes in the incentive program favoring investments in the north have resulted in increase plantation activities with preliminary results reported as favorable. Activities in this region involve largely tropical pine. Finally, in the Amazon, there is the famous Jari project of Daniel Ludwig. To date, Jari has some 100,000 hectares of forest. Initially, the plantings were almost exclusive gmelina (Gmelina arborea), but more recently tropical pine (P. caribaea) has been favored on the poorer sites. Current plans in Jari call for the introduction of eucalyptus (deglupta) on the intermediate sites. Substantial additional plantings of eucalyptus, pine and gmelina are envisionsed over the next several years. For pine, growth rates of 18 cubic meters per hectare per year are probably representative while somewhat more apid growth can be expected for the eucalyptus and gmelina. Should urrent plans be realized, some 400,000 hectares would eventually become lantation forests. Elsewhere in the Amazon the introduction of plantation orests is moving slowly as the government has opted for greater efforts to tilize and manage the indigeneous forest.

Potential Future Production from Brazilian Plantations

What are the implications of the recent forest plantation activities f Brazil upon world markets? The previous section has outlined briefly ome of the dimensions of the recent and planned forest plantation activi- ies in Brazil. This section assesses the potential Brazilian wood availa- lity in the mid-1990s. The analysis suggests that under a variety of cir- umstances Brazil is likely to have a wood inventory of sufficient size to ave the potential to dramatically impact upon the world market for wood lp and, particularly, chemical wood pulp.

cent Trade Performance

Brazil has not historically been a major trader of forest products. the 1960s Brazil did generate a small export surplus in solid wood pro- cts such as sawnwood, and panels. However, the trade surplus has genera- y been declining as domestic consumption utilizes greater portions of do- stic solid wood production. In paper products, Brazil has experienced a dest trade deficit in recent years. In general, the forest product lance of trade has experienced a gradual deterioration in the past decade.

An exception to this trend is wood pulp and, particularly, chemical od pulps. During the 1966-77 period, Brazilian production of chemical od pulps grew at almost 11.5 percent annually, while the worldwide rate was only 3.5 percent. For all wood pulps, Brazil moved from a trade deficit of 16,000 tons in 1966 to a modest surplus of 20,000 tons in 1976 (table 4). This trade balance improvement was due to chemical wood pulp which grew from a modest trade surplus of 3,000 tons in 1966 to 39,000 ton in 1977 (FAO, 1979).

Taken by itself, the improvement in Brazilian production and trade position for pulps and chemical pulps might simply be interpreted as an obscure curiosity. However, in light of the massive conversion of Brazilian lands into forest plantations, the extraordinary biological growth rates that are achieved, and the massive additions to plantation lands that are planned for the early 1980s, it becomes apparent that the potential impact upon world markets of the additional production of pulp available for world markets in the 1990s from Brazil could well be large.

Quantitative Impacts

Given these considerations it is instructive to speculate as to the potential quantitative impact that forest plantation activities in Brazil may have on world pulp markets by the mid-1990s. As noted earlier, Brazil at the end of 1978 had about 3.5 million hectares in plantation forests, and with the estimated addition of 0.3 million hectares in 1979 the current total is about 3.8 million hectares. Of this, perhaps 2.5 million hectares is eucalyptus, with the remaining 1.4 million hectares largely in pine. The average productivity of these forests varies from about 12.5 cubic meters to 35.5 cubic meters per hectare per year. A reasonable average might be about 20 cubic meters for pine and 25 cubic meters per hectare per year for eucalyptus. Using these averages, the current plantations are producing a total of about 90.5 million cubic meters of pulpwood per year, of

Table 4.   Brazil's Wood Pulp Production and Trade, 1977

(million tons)

	(1) World Total 1977	(2) Brazil 1977	Brazil as percentage of world total
Wood pulp production	116	1.6	1.4
Chemical wood pulp production	78	1.3	1.7
International trade: Wood pulp	17	+.02[a]	0.3
International trade: Chemical wood pulps	14	+.04[a]	0.6

Source: Yearbook of Forest Products: 1977 (Rome, FAO).

[a]Brazil's net exports.

which about 62.5 million cubic meters are eucalyptus and 28 million cubic
meters are pine. As noted above, current Brazilian plans call for an addition of about 0.5 million hectares per year through 1985, reaching a total
of about 6.8 million hectares. If these plans should be realized and the
1978 planting composition of 1966-78 between pine and eucalyptus be re-
tained, the result would be about 4.3 million hectares in plantation eucalyptus and 2.5 million hectares in pine by the mid-1980s. Given current
productivity levels, Brazilian plantations would then be producing 107.5
million cubic meters of eucalyptus and 50 million cubic meters per year of
pine for a total of about 157.7 million cubic meters of wood per year.

According to the FAO, world total production of all wood pulp in 1977
was 116 million tons, of which only 1.4 percent was Brazilian production,
while world chemical pulp production in 1977 was about 78 million tons, of

which Brazil produced about 1.7 percent. An upper limit to Brazilian potential production can be estimated if it is assumed that all the wood from
forests expected to be planted through 1985 and available for harvest in th
1990s would be converted into pulp (at the conservative factor of 5 cubic
meters of wood per ton of pulp). In this case, Brazil would be capable of
providing a sustained flow of wood from her plantations equal to the wood
inputs required to produce 31.5 million tons of pulp annually. Utilizing
the FAO projections for 1995, 31.5 million tons of pulp would equal 17 percent of total FAO projected worldwide production of all wood pulps and abou
26 percent of FAO projected chemical pulp production (FAO, n.d.).[5]

The above illustration is clearly an upper limit estimate of potential
Brazilian pulp production in mid-1990s (but not for the early twenty-first
century if continued new plantings are forthcoming beyond 1985), since all
of the sustainable wood harvest is assumed to be directed into chemical
pulps. Let us explore some other, more realistic possibilities.

High Scenario

As previously noted, Brazil does have uses for her plantation timber
besides pulpwood. An important use is charcoal. The use of wood for charcoal for the Brazilian steel industry has been well established and is undoubtedly going to expand. If 80 percent of the current plantation production in Minas Gerais State is directed to charcoal production, as has been
estimated, then about a third of total Brazilian eucalyptus plantation

---

[5]The FAO (n.d.) study used an implicit growth rate of 2.8 percent per
annum for world chemical pulp production. This factor was applied to the
1990 FAO projection to give a projection for 1995, a year not projected in
the FAO study.

acreage of 1978 can be viewed as being utilized for charcoal production. If this relation holds for new plantations created through 1985, about 1.4 million hectares of eucalyptus lands can be viewed as being diverted to charcoal feedstock production (35 million cubic meters of wood per annum). This implies that the production of the remaining plantations, 122.5 million cubic meters, would be potentially available for pulp and could produce 24.5 million tons of pulp.

To assess the potential impact upon trade flows, an adjustment must also be made for domestic consumption. If Brazil's domestic consumption of pulp increases to 4 million tons by the mid-1990s, the total amount potentially available for export, that is, excess supply, would equal about 20.5 million tons.[6] This is about 60 percent of 1995 projected world pulp trade and 73 percent of projected chemical pulp trade (this is the high scenario of table 5).

An additional dimension of realism can be added by considering the solid wood potential of Brazilian forests. If we assume that 20 percent of the potential total volume of the noncharcoal forest production (124.5 million cubic meters) is utilized as some form of solid wood, a reasonable assumption given thinnings, residuals, and the like, there remains about 98 million cubic meters of wood residuals and pulpwood that can produce about 20 million tons of pulp, or 14 percent of 1995 projected total world production of chemical pulps. After domestic needs are met, there would remain 16 million tons for the world market. This is equal to 46 percent of projected total international trade in chemical pulps for 1995 and about 17 percent of total world production.

---

[6]The projection assumes Brazilian domestic consumption of all pulps rows at about 5.4 percent per year from a base of 1.45 million tons in 1976.

Low Scenario

The limiting case of potential on the conservative side assumes that Brazil ceases new plantation creations after 1979, a highly unlikely prospect. In this event, the 1995 potential would be reduced to about 50 percent of our preceding scenario. In this case, utilizing our assumption about drawdowns for charcoal and solid wood, the total chemical pulp availability could be approximately 10 million tons, and allow for exports of 6 million tons after meeting domestic demand. While the fraction of world trade is considerably below the levels discussed above, Brazil's potential pulp exports will amount to over 21 percent of projected 1995 intenational trade for all pulps and 25 percent of projected chemical pulp trade. Again, Brazil's impact on the world market could be large (see table 5).

Long-fiber Versus Short-fiber Pulp

The discussion thus far has treated chemical pulps as homogeneous and has not recognized the important differences between the long- and short-fiber chemical pulps and the differences that exist in the markets for these two commodities. It can certainly be argued that the impact upon the United States of forestry of plantations elsewhere in the world is dependent upon the type of forest resource that is being produced. For the United States and North America, the vast majority of the forest products are from conifers and therefore are designated as softwood or long-fiber. Extending our Brazilian analysis one step further, the chemical pulps can be crudely disaggregated to determine the potential impacts upon these two somewhat different markets.

Table 5. Brazil's Potential Wood Pulp Production and Trade, 1995

(million tons)

	Projected world totals 1995[a]	Brazilian potential 1995[b]			
		High Scenario	World Levels (%)	Low Scenario	World Levels (%)
Wood pulp production	190	24.5	13	10	5
Chemical pulp production	120	24.5	20	10	8
International trade: All wood pulps	34	20.5	60	6	18
International trade: chemical wood pulps	28	20.5	73	6	21

[a]World production for 1995 based upon FAO World Outlook: Phase IV World Outlook for Fiber Products, Prepared by a Joint Forestry and Industry Working Party for the Forestry Department of the FAO (Rome, FAO, 1979).

[b]Brazilian projections by the author.

After withdrawing from consideration the eucalyptus production designated for charcoal, 41 percent of the remaining projected 1995 production
will be long-fiber. FAO projections suggest that world chemical pulp production will consist of 29 percent short-fiber by 1995. For our high scen
ario, the implication is that Brazilian potential exports in 1995 could
constitute as much as 156 percent of the world's trade in short-fiber chem
ical pulps and 40 percent of the world total trade in long-fiber chemical
pulps. For the low scenario, the percentages fall to 44 percent of the
short-fiber pulp and 12 percent of the long-fiber total world chemical pul
trade. Although impact is greatest on short-fiber pulps, the potential im
pact on long-fiber pulps continues to be significant.

Implications for the United States

What are the implications for the foregoing analysis of the potential
impact of Brazilian plantation forest production on the future role of the
United States in world forest trade? The implications are of two types--
specific and general. The specific implications relate to the impact of
potential Brazilian exports upon world markets for chemical pulp and the
ramifications upon world market trade and U.S. producers. As the analysis
shows, in as short a time as fifteen years, the impact of Brazilian production can, under a variety of assumptions, have a potentially large effe
on a particular world commodity market. There are many qualifications to
the above analysis of Brazilian potential. Wood on the stump is not suffi
cient to generate exports of pulp. Obviously, additional pulp mills must
be forthcoming and transportation networks must be in place. In addition,
financial, or other incentives, must be present to justify the harvesting

and processing activities. Therefore, the above projections are not intended to be a forecast in the usual sense. Rather, the intent of the analysis is to provide an "order of magnitude" quantitative estimate of the potential for pulp production that now exists in Brazil, based on current activities, near-term plans, and existing practices.

However, the broad general implications of the analysis go well beyond the estimation of the potential impact of a particular "case study" upon a commodity market. As was indicated earlier in this paper, the forest resource is experiencing an important transition from natural to plantation forest, and this process has been gaining momentum. Associated with the transition to plantation forestry is the possibility of developing important forestry resources in countries that have not been traditional producers and exporters of large volumes of forest products. As noted earlier, plantation forests are being created in numerous regions of the globe that have not been traditional exporters. Although most of these regions will not become major forest product exporters, some almost certainly will. Thus new and important supply sources are likely to be added to the traditional suppliers.

The discussion is more than hypothetical as the process is already well along. The exotic forest resources of New Zealand, Chile, and Brazil are beginning to find their way into world markets. Although the process may be quite gradual, this paper suggests that for some commodities, for example, short-fiber chemical pulps, the change could occur rather dramatically. In addition, both Chile and New Zealand have substantial stands of mature or near mature conifer that could enter world markets either as long-fiber or as solid wood. Both of these countries are likely to increase their exports very substantially in the next ten or twenty years.

In recent years, U.S. awareness of international forest resources and forest products trade has increased dramatically. However, the view is still largely in terms of the impact that the United States can have on world markets either as an exporter or an importer. The usual approach is to project domestic demand and supply, with perhaps an integrated Canadian component. Imports and exports are treated as exogeneous and projected largely on the basis of past trends and traditional supply sources. However, this paper suggests that forces external to the United States are likely to have profound impacts upon the structure of world trade in fores products. These forces could ultimately impact forest industry investment decisions, future land use patterns, the nature of forest product inter-country competition, and the role of the United States and North America in world forest products trade.

REFERENCES

Beattie, William David, 1975. "An Economic Analysis of the Brazilian Fiscal Incentives for Reforestation," (Ph.D. dissertation, Purdue University).

Carbonnier, Louis, and Goran Lonner, 1976. "Aracruz: Forestry Creates a Business--How Foresters Aim for Profitability," World Wood (June 1976) pp. 24-26.

Instituto de Pesquisas e Estudo Florestais, n.d. "Foresta en quantidade e qualidade para desenvolvimento economico e bem-estar social," (IPEF).

Johnson, Norman E., 1976. "Biological Opportunities and Risks Associated with Fast-Growing Plantations in the Tropics," Journal of Forestry vol. 75, no. 4 (April) pp. 206-211.

Kellison, Robert C., 1979. "Advances in Tropical Tree Improvement." Paper presented at the International Symposium on Tropical Forests and Their Contribution to the Development of Tropical America, San Jose, Costa Rica, October.

Ministerio da Agricultural, IBDF, 1978. Dianostico da Participacao do Sub-setor Florestal na Economia Brasileria (Brazil).

U.N., Food and Agriculture Organization, 1979. Yearbook of Forest Products: 1966-1977 (Rome: FAO, 1979).

_____, n.d. "FAO World Outlook: Phase IV." (Rome, FAO) pp. 24-28.

Zobel, Bruce, 1979. "Timber Supply Trends in South America." Paper presented at CICPELA, Cartagena, Columbia, November 5.

Discussion by Jan G. Laarman

Roger Sedjo counsels us to look south, not just north, to take account
of important supply regions impacting on world forest products trade in
the next decade or two. His paper concentrates on Brazil. To fall back on
a cliche, Brazil has been a sleeping giant that is now awakening. Or is
it the sleeping United States which is now awakening to the very real presence of a major wood-producing area growing up outside of North America?
And does Brazil's potential in the forest products arena signal a more general awakening that south is not merely a direction on the compass, but
also a possible direction to be taken by the world's changing economic
order?

Most of us have been aware that Brazil is a big country which has sizable areas of fast-growing forest plantations. However, we usually lack
any kind of quantitative framework within which to assess their empirical
implications. Sedjo's paper constructs one such framework. It lays out
alternative scenarios of Brazil's potential fiber production and exportable surplus, and then measures this surplus with a world yardstick. When
scaled in this manner, Brazil's likely impact on world fiber trade is indeed impressive.

However, Sedjo is very clear to state that wood on the stump is not
sufficient to generate exports of pulp. That is, he carefully distin-

uishes physical stocks of wood from economic supply. He also makes the
ualification that his projections are not the usual kind of _forecast_, but
re rather to make us cognizant of Brazil's wood-producing _potential_.

My comments will not take issue with Brazil's planting statistics,
iological growth rates, regional distribution of species, or other ele-
ents of plantation geography. I have no grounds to dispute a few tacit
ssumptions: that Brazil's sociopolitical order will not collapse; that
razil will be able to cope with domestic problems with high rates of mone-
ary inflation and large external debt; and that Brazil's plantations will
ot be lost to pathogens.

A more tangible question is energy, a subject of pervasive concern to
razil. After years of costly exploration, Petrobras has not been able to
ocate petroleum reserves of any consequence. This partially explains why
razil has embarked on the world's most ambitious program to produce al-
hol from biomass. Current efforts are directed to bagasse (that is,
ugar cane residue), but the technology could lead elsewhere. In partic-
lar, it could lead to plantation-grown wood.

Because Brazil is apparently on the leading edge of applying these
echniques, it may emerge as the world's pioneer to obtain energy from
ood on a large scale. That possibility is underlined by Brazil's recog-
ized expertise in growing eucalyptus for industrial charcoal. If wood
or energy (and chemicals) looks feasible in the years ahead, then fiber
nd energy will compete as end uses to an extent not previously contem-
lated.

Still, I concur with Sedjo's thesis that Brazil will most probably
hange the structure of world pulp and paper trade to a major degree. I

will argue, however, that few other "new" countries are likely to enter
the export picture with impacts on the United States comparable to those
we expect from Brazil. Even then, Brazil's fiber exports seem likely to b
heavily concentrated in hardwoods, not softwoods, for several reasons not
explicitly cited by Sedjo.

Malthus Revisited

A case study of Brazil leaves us with little latitude for general inferences. Brazil's plantations are distinguished not only for their sheer
scale, but also for their private ownership. This is as opposed to goverm
mental and quasi-governmental plantations in many other countries, where
world market signals may not be reached as clearly. Even more important i
that Brazil's plantations were established without the food shortages so
obvious in Asia and Africa, where several countries contend with Malthusia
ratios of population growth to arable land.

With a few exceptions we shall enumerate, exotic plantations in most
new potential supply regions fit one or both of these categories: (1) cor
pensatory for the depletion of natural forests, or (2) planted for reasons
other than to produce industrial timber. A few small countries with small
populations will have plantation timber or products to export in the near
future. An example is Fiji. However, the number of new supplying countries to have export capacity on a significant scale (from the U.S. perspective) comprises a short list. Besides, Brazil, the most obvious countries are New Zealand, Chile, and South Africa. In a later time frame, we
may also look at Venezuela. Interestingly, only Brazil and Venezuela have
land lying within the tropics. Yet even many of Brazil's plantations--
including the principal areas of pine--are not tropical.

Apart from the five countries cited in the preceding paragraph, most other new supply regions will be limited by intense competition for the services of productive land. Especially in the developing countries of Asia and Africa, the obtention of industrial wood is often mixed with objectives of providing watershed protection; organic fertilization; fuelwood; livestock fodder; oils, barks, dyes, and extractives; and fruits, seeds, and nuts for human consumption. Thus many plantations follow any of various schemes to combine tree crops with food crops or animal grazing. An "agro-forestry" approach seems eminently sensible for heavily populated, land-poor countries (Asian Development Bank, 1978, pp. 44-48). However, it will necessarily constrain the flow of industrial roundwood from these lands, and therefore also the likelihood of achieving exportable surplus.

A recent FAO publication by Lanly and Clement (1979) projects the decreasing area of natural forests and the increasing area of plantation forests for subgroups of ninety seven developing countries to the year 2000. The study depends heavily on guesswork and data of dubious reliability. Yet it gives us a rough idea of general expectations. I will use a few figures from that study to suggest that plantations in developing countries will not furnish a cornucopia of industrial wood. On the contrary, the outlook is somewhat grim.

Most of the countries studied by Lanly and Clement are at least partly tropical or subtropical. They exclude South Africa, China, and a few other countries with important plantation areas. For the ninety seven countries covered, Lanly and Clement project that the area under industrial plantations will more than double from 9,095 hectares in the year 1980, to 21,150

hectares in the year 2000.[1] The largest rate of increase is projected
for Latin America, followed by Asia and the Far East. The smallest rate
of increase is projected for Africa. By the year 2000, Latin America
(mainly Brazil) will have just over half of all industrial plantations
in the ninety seven countries.

For this group of ninety seven, an increase in industrial plantations
amounting to more than 10 million hectares has to be veiwed against a depletion of closed tropical forests. Lanly and Clement suggest that this
depletion will amount to 140 million hectares in the twenty five years
from 1975 to 2000. But cutting in tropical forests, whether for land
clearing or other purposes, often yields only small volumes of industrial
wood. Hence a simple comparison of plantation area gained with natural
forest area lost is grossly misleading. But not all plantations are high-
yielding, either, and this point deserves emphasis.

Lanly and Clement show that in 1975, the ratio of nonindustrial to
industrial plantations was .29 in Latin America; .58 in Africa; and 1.32
in developing Asia and the Far East. Also, a relatively large proportion
of the industrial plantations were "low yielding," that is, having mean
annual increments at rotation age of less than 15 cubic meters per hectare
per year. Low yielding industrial plantations as a proportion of total
industrial plantations were .25 in Latin America; .70 in Africa; and .84
in developing Asia and the Far East.

---

[1]Lanly and Clement define industrial plantations as those established
at least partly for the production of industrial wood in the form of sawtimber, pulpwood, or mining props. Nonindustrial plantations are aimed
entirely at other uses, for example, fuelwood, fruit production, and so on.

Low-yielding plantations in Africa and Asia include large areas of slow-growing hardwoods, for example, cabinetwood species like teak and mahoganies. That is, plantations can be low-yielding because of species, not just low land productivity. With respect to impacts on the United States, quality tropical and subtropical hardwoods often do not have close substitutes, and are not "commodity woods" in the ordinary sense. Rather, they enter a special niche in world trade not likely to shift investment patterns or market shares in new directions, except among the producer countries themselves.

Lanly and Clement (1979) do not cite individual countries, except in larger regional groups. An earlier study by Persson (1974) does focus on countries, and we can use Persson's data (even if somewhat old) to identify those countries which had the largest areas planted as of the early 1900s. This will help assure that we do not overlook single countries of special significance.

According to Persson, the regions with the largest areas of forest plantations were: China, 20,000+ hectares; North America, 10,600 hectares; Japan, 8,900 hectares; and Europe and USSR, 7,500 hectares. None of these regions fits Sedjo's concept of a "new" supply region likely to have near-term exportable surplus from exotic species. Despite the vast area of plantations in China, it epitomizes the wood-deficient, heavily populated country pursuing an agro-forestry policy (Richardson, 1966).

On the other hand, forest plantations in Japan could obviously have profound impacts on world trade. The effect would likely be one of import substitution rather than one of exportable surplus. An overview of forest plantations in Japan has been published recently (Matsui, 1980). We may

infer from that article that Japan could begin to tap its softwood plantations on a greatly expanded scale before the year 2000. Depending on
the growth of Japanese demand, that could cause downward adjustments for a
number of present and anticipated exporting countries, and would certainly
affect U.S. forest products trade both directly and indirectly. But because Sedjo's objective is to raise the issue of new countries to enter
the trading mainstream we will not say more about well-entrenched Japan.

Persson's second tier of plantation regions comprises Asia (excluding
China and Japan), 5,400; Africa, 2,800; Latin America, 2,700; and Australia
and New Zealand, 900. These are the groups we need to examine more carefully.

With respect to Asia, Persson shows that the countries Republic of
(South) Korea, Indonesia, and India accounted for 80 percent of the total.
The Asian Development Bank classifies India and the Republic of Korea as
"countries deficient in forest resources, but at a high level of industrial
development" (Asian Development Bank, 1978, p. 9). We could therefore expect domestic consumption to absorb most of plantation output. As for Indonesia, different parts of that country exhibit widely varying forest
area per capita. Plantation wood in densely populated Java and Sumatra
may be consumed for local use, not export.

In Africa, South Africa was the only country to have more than a million hectares (1,025,000) of forest plantations in the period 1969-72.
Other leaders were Morocco (294,000), Madagascar (240,000), Kenya (138,000)
and Tunisia (114,000). FAO's Yearbook of Forest Products (1979) shows
that all of these countries--including South Africa--have been consistent
net importers of forest products on a value basis. For 1977 the ratios of

mports (c.i.f.) to exports (f.o.b.) were 2.3 for South Africa; 9.9 for
orocco; 4.6 for Madagascar; 6.4 for Kenya; and 21.3 for Tunisia. Together
hese five countries accounted for only 0.3 percent of world forest product
xports.

For Latin America, Persson's statistics show Brazil with 1,350,000
ectares (year 1972), although Sedjo's tables indicate that by 1978 this
ad increased at least 115 percent. Behind Brazil were Chile (440,000),
rgentina (325,000), Cuba (215,000), and Uruguay (155,000). Each of the
hree latter countries is a net importer of forest products at the present
ime, and of them, only Argentina has exports of significance.

Chile, however, has been a net exporter by a large margin (16:1 in
977). In the period 1966-77, Chile's value of exports grew at a rate al-
ost double the world average (FAO, 1979). In 1977 pulp and paper com-
rised more than 80 percent of the value of Chile's forest products ex-
orts. But Chile has a problem with its age-class distribution, so that
 gap exists between its older plantations and those just recently estab-
ished (Zobel, 1979, p. 40). Hence production and exports could conceiv-
ly fall in the period after the older plantations have been liquidated.

According to Zobel (1979, pp. 39-40), another country which has great
otential is Venezuela. Large areas of flat lands are suitable for estab-
ishing pines, and the government is interested in planting. But major
lanting began only recently, and planting practices need to be improved.
enezuela may not have pulp to export for fifteen to twenty years. The
hree countries--Brazil, Chile, and Venezuela--are the only Latin American
roducers which Zobel sees as having other than very limited supply impacts
 the United States.

Finally, let us consider the Pacific Islands, principally Australia and New Zealand. Australia is an industrialized country with "adequate" forest resources, according to the Asian Development Bank (1978, pp. 31-32). It is struggling for self-sufficiency in forest products, but at high cost because of its small and scattered markets.

In contrast, New Zealand's export capacity is unquestioned. Ratios of export value to import value were 1.7 in 1966, 5.1 in 1971, but 4.7 in 1977 (FAO, 1979). Exports are expected to increase after 1985 to reflect an inventory buildup explained by the expanded planting which began in the early 1960s. It is still uncertain how exports will be divided between solid wood and fiber. Another unknown is how the rising costs of fossil fuels will impact on ocean freight rates, and hence New Zealand's competitiveness in markets as distant as Europe.[2]

The preceding observations call attention to our need for a comprehensive and updated timber trends study at the world level, which would give special attention to plantations. However, until such data come along, I will argue that the important countries to have plantation-wood exports are few in number. Even some of the shining starts may have their specks of tarnish: Chile, unbalanced age-class distribution; South Africa, explosive politics; New Zealand, costly transport. The transport conundrum clearly affects the other Southern Hemisphere exporters, as well. Some of the obstacles faced by Brazil, particularly with respect to generating exportable surplus from its pines, are discussed in the following section.

---

[2]Personal communication, Russ Ballard, Department of Forestry, North Carolina State University, Raleigh, North Carolina.

Production from Brazil's Pine Plantations: How Much Will Be Exported?

Sedjo indicates that Brazil's share of world exports will be substantially higher for hardwoods fiber than for softwoods fiber. But the reasons for this go beyond merely comparative areas planted and comparative volumes of world trade, hardwoods fiber versus softwoods fiber. Large-scale exports of Brazilian hardwoods fiber are already under way. Yet the exportable surplus to be derived from the nontropical pine plantations located south of São Paulo may not materialize to any significant extent. I will attempt to explain why a conservative outlook is warranted.

Sedjo's table 3 shows the pattern of pine plantation establishment by state from 1967 to 1978. Although much of the recent planting has shifted northward to Minas Gerais, Mato Grosso, and Bahia, the three southern states—Paraná, Santa Catarina, and Rio Grande do Sul—manifest 60 percent of the Brazilian total through 1978. Plantations in these southern states are principally P. taeda and, secondarily, P. elliotti.

With respect to these "southern pines," it is still too early to estimate the relative volume and number of stems that will be consumed in the form of sawtimber, veneer bolts, posts and poles, and pulp wood. However, there should be considerable pressure to utilize plantation pine for sawtimber and veneer.

The three southern states historically produced most of Brazil's solid wood. Although the araucaria timber is essentially cut over, the entrepreneurial base remains one of predominantly small firms oriented to production of sawnwood and veneer. If the domestic demand for solid wood does not abate, and this seems a reasonable assumption, the feedstock for fiber production may be little more than thinnings and by-product chips.

We could speculate that native hardwoods in Mato Grosso and elsewhere, which in part now substitute for solid wood araucaria, will in turn be substituted by plantation pine. If pulpwood is accorded no more status than residual use, its effective volume could be relatively minimal.

Some of the sawmillers in the Brazilian wood-processing industry are thinking in terms of 20-centimeter diameters as an acceptable lower limit for pine plantation sawlogs. Studies have shown that 60 to 70 percent of the trees in unthinned plantations exceed 20 centimeters at ages twelve to fourteen.[3] These findings seem to suggest that only the volumes from the first thinning (ages seven to eight) and some proportion of the second thinning (ages ten to twelve) will become available as pulpwood. The largest proportion of the volume from the third and subsequent thinnings would eventuate in non pulpwood uses. It is emphasized that this is a purely speculative view, one which remains to be tested against future practice. In any event, the emerging pattern by which plantation pine is allocated between solid wood and fiber will be closely watched both inside and outside of Brazil.

Another limitation for pulp production is that fast-grown pines show reduced fiber quality and yield. Fast growth means that the trees become merchantable when still young, and therefore contain a high proportion of juvenile wood. As a consequence, solid products have weaknesses associated with wide rings. Pulp yields are low (per units of roundwood input) and have low tear strength. Studies confirm that the costs of pulpmill operations are increased when the content of juvenile wood is increased (Semke

---

[3]Measurements in IBDF's _florestas_ _nacionais_ at Irati and Capao Bonito.

and Corbi, 1974). These factors could lessen Brazil's international competitiveness.

Also, the pine plantations are generally small and dispersed. The
setting of small-scale sawmill entrepreneurs and comparative small land
holdings spilled over into small and fragmented reforestation projects.
A mid-1970s survey of plantations in south-central Paraná found that the
vast majority of plantings were less than 200 hectares in size. The single
largest plantation was 1,960 hectares (Consultoria Brasileira Florestal
Ltda., 1974).

A handful of established pulp and paper mills will be able to draw
upon captive plantations. However, any new enterprise opting for large-
scale fiber export will have to (1) purchase a large number of separate
plantation holdings from a large number of different owners; or (2) acquire
a land base on which to establish its own captive plantations. Either
alternative will be costly.

Also bearing on local costs, and ultimately on international competitiveness, is plantation-wood transport. Outside of the southern states, a
few mammoth projects like Jari and Aracruz were planned around water transport, at least for product shipment. However, the southern states' pine
probably will be transported mainly by truck. The low density of railroads
and rivers does not afford many opportunities for transshipment. In late
1974, truck freight rates in southern Brazil accounted for 30 to 40 percent of the costs of pulpwood delivered at the mill, and typical one-way
hauling distances were 80 to 400 kilometers.[4]

---

[4]These figures are from Industrias Klabin do Paraná S/A, Monte Alegre,
Paraná.

The preceding arguments are intended to convey a cautious view of exportable surplus from Brazil's pine plantations. However, I would like to conclude with a key counterpoint not fully developed by Sedjo. It is that Brazil's plantations--both pines and hardwoods--are currently growing well below their biological potential. Zobel (1979, p. 38) attributes this to excessive haste to expand the area planted without due regard for proper selection of sites, supervision of planting practices, and use of quality seed and planting stock. Thus, what is now "fast growth" from the U.S. perspective will probably become even more impressive in the future. In this regard, a parting tribute may be: We ain't seen nuthin' yet.

53

REFERENCES

sian Development Bank. 1978. _Role of the Bank in Forestry and Forest Industries Development_ (Manila, ADB).

onsultoria Brasileira Florestal Ltda. 1974. _Levantamento dos Reflorestamentos Existentes na Regiao de Guarapuava_ (Curitiba, Brazil, CONF-AL).

ood and Agricultural Organization of the United Nations. 1979. _Yearbook of Forest Products 1966-1977_ (Rome, FAO).

anly, J.P., and J. Clement. 1979. _Present and Future Forest and Plantation Areas in the Tropics_, FO:MISC/79/1 (Rome, FAO).

atsui, Mitsuma. 1980. "Forest Resources in Japan," _Journal of Forestry_ vol. 78, no. 2, pp. 96-99.

ersson, Reidar. 1974. _World Forest Resources_ (Stockholm, Department of Forest Survey, Royal College of Forestry).

ichardson, S. D. 1966. _Forestry in Communist China_ (Baltimore, Md., Johns Hopkins University Press).

emke, K., and J. C. Corbi. 1974. "Sources of Less Course Pine Fiber For Southern Bleached Printing Papers," in _Proceedings of TAPII Meetings_ (Miami).

obel, Bruce J. 1979. "The Southeast Timber Supply: How Will It Be Affected by Changing Forestry in South America, Canada, and the Northeast?" _Southern Journal of Applied Forestry_ vol. 3, no. 2, pp. 37-42.

U.S. EXPORTS AND IMPORTS OF SOME MAJOR FOREST PRODUCTS--

THE NEXT FIFTY YEARS

David R. Darr

Projections of long-term U.S. supplies and demands for timber products have been with us for the past one hundred years or so. Current U.
Forest Service efforts in this area stem from the Forest and Rangeland Re-
newable Resources Planning Act of 1974 (act of August 17, 1974--Stat. 476
as amended; 16 U.S.C. 1600-1614). This legislation mandates a periodic
Renewal Resource Assessment and includes, but is not limited to, "an analysis of present and anticipated uses, demand for, and supply of the re-
newable resources, with consideration of the international resource situation, and an emphasis of pertinent supply and demand price relationship
trends." For the assessment of the forest and rangeland situation in the
United States (U.S. Forest Service, 1979), the Foreign Trade Analysis
Work Unit in Portland, Oregon, was given primary responsibility for developing projections of long-term trade in timber products. The entire work
unit contributed to the development of the projections. In this paper, I
will report the results of the work unit's efforts. The projections were
used to interpret how international markets might affect U.S. supplies of
and demands for, timber products. In this case, long term means just

that--projections of both U.S. domestic markets and trade were made to the year 2030, a fifty-year period. A period of this length is considered essential in order to take actions now to offset any undesirable market situations that might develop over the coming decades.

The primary purpose of this paper is to present the projections of U.S. imports and exports of major timber products used in the 1979 assessment, including discussions of the rationales for the projections. I will also discuss the methodology used in making fifty-year projections. From a conceptual standpoint, U.S. trade patterns are influenced by the trade patterns of other countries and vice versa. For example, what happens in Japan affects what happens in the United States which affects what happens in Canada which affects what happens in Europe, and so forth. The following points were especially important in assessing prospects for U.S. imports:

1. The size and other characteristics of the U.S. market. A growing, dynamic domestic market would lead to different demands for U.S. imports than would a stable or declining market.

2. The supply situation in source countries. The availability of supplies in traditional source countries for U.S. imports has the potential to affect the price and quantity of imports from these countries. If supplies are expanding or at least available, this would present a much different prospect for imports than if supplies are declining in volume or quality, or both.

3. Competing demands from other consuming countries. Import prospects for the United States would differ when the United States had no competition in purchasing imports from source countries than when there was competition for available supplies.

4. The competitiveness of U.S. producers. If U.S. producers of a timber product are efficient, agressive, and cost-competitive with producers in import source countries, a much different prospect for imports is presented than if the U.S. segment of the industry has trouble remaining competitive.

The following points were especially important in assessing prospects for U.S. exports:

1. The expected size and other characteristics of the U.S. domestic markets. For example, if the U.S. market for a product is expected to be one of cost-competitive producers with limited demand, this would lead to a different assessment of export prospects than if domestic demand is expected to grow rapidly.

2. The demand situation in export market areas. For example, if demands are expected to grow in a traditional U.S. export market, this would present a different assessment of export prospects than if demands are expected to be stable or decline.

3. The situation in competing supply areas. For example, if the competition for export markets is expected to be strong, this would lead to a different assessment of U.S. export prospects than if the competition is expected to be weak.

Thus, in our fifty-year projections of U.S. trade in timber products, we have attempted to account for major market interactions that might affect trade patterns.

In making our projections, we had limited time, resources, and manpower. Thus, we set priorities and limited any detailed analysis to products and countries that appear to have the most potential to affect U.S. trade patterns. Between the early 1960s and the mid-1970s, lumber imports (almost all softwoods from Canada) increased in relative importance in the U.S. product mix of imports, and newsprint declined in relative importance (table 1). The categories of lumber, pulp, newsprint, and other paper and board products account for well over three-fourths of the value of U.S. imports of timber products. Almost all U.S. imports of these products originate in Canada.

Although veneer and plywood imports account for less than 10 percent of the value of U.S. imports, they account for about two-thirds of total U.S. consumption of hardwood veneer and plywood. Most of these imports originate in South Korea, Taiwan, the Philippines, and Singapore. The logs for a major share of these veneer and plywood shipments, however, originate in Indonesia and Malaysia.

Table 1. Imports by Commodity as a Percentage of the Value of all U.S. Imports of Timber Products, 1960-62 and 1975-77

Commodity	1960-62	1975-77
Logs and pulpwood	1.5	0.9
Lumber	19.1	25.0
Pulp	20.3	20.7
Newsprint	41.1	31.0
Other paper and board products	4.7	5.8
Veneer and plywood	6.6	8.1
Other	6.7	8.5
Total	100.0	100.0

Source: U.S. Department of Commerce (1978b), U.S. Imports: Commodity by Country, FT-410 (Washington, D.C.).

Table 2. Exports by Commodity as a Percentage of the Value of all U.S. Exports of Timber Products, 1960-62 and 1975-77

Commodity	1960-62	1975-77
Logs and pulpwood	7.2	22.9
Lumber	14.7	11.4
Veneer and plywood	0.9	3.7
Pulp	26.4	19.3
Newsprint	2.9	1.0
Other paper and board products	40.1	32.2
Other	7.8	9.5
Total	100.0	100.0

Source: U.S. Department of Commerce (1978a), U.S. Exports: Commodity by Country, FT-410 (Washington, D.C.)

Imports of other timber products are generally inconsequential in judging prospects for U.S. supply and demand.

In our projections of U.S. imports of timber products, we gave detailed attention to prospects for lumber, pulp, newsprint, and other paper and board imports from Canada and to prospects for hardwood veneer and plywood imports from southeast Asia.

Between the early 1960s and the 1975-77 period, the value of exports of logs and pulpwood gained in importance relative to other product categories (table 2). Other major categories in the product mix of U.S. exports are lumber, wood pulp, and other paper and board, especially linerboard. Although veneer and plywood exports account for less than 4 percent of total U.S. sales, there has been much interest in recent years in

he export market potential for softwood plywood. In our projections of
.S. exports of timber products, we gave detailed attention to prospects
or logs, pulpwood, lumber, veneer and plywood, wood pulp, and other paper
nd board products. Exports of other products were considered inconsequen-
ial in the U.S. timber supply-demand situation. Detailed attention was
iven to Japan and western Europe in assessing export prospects. As dis-
ussed later, exports to other areas have limited potential for growth.

Softwood Logs

In evaluating U.S. timber supplies and demands, we expect imports of
oftwood logs to remain inconsequential. Imports are projected to remain
: 100 million board-feet throughout the projection period (table 3).
anada's long-standing restrictions on log exports will keep volumes at
istorical levels. Most of the interest in U.S. log trade has centered on
oftwood sales to Japan, which account for over 80 percent of U.S. exports.
xports of softwood logs increased from 600 million board-feet in 1962 to
ell over 3.5 billion board-feet per year in the 1970s. We project U.S.
xport volume to stay near current levels until 1990 and then to decline
:adually to a still significant 2.5 billion board-feet in 2030.

The decline in volume after 1990 reflects our expectations about the
et effect of several interacting market forces. Available timber supply
:ojections for the West Coast indicate a decline in harvest on industry
inds, the source of over two-thirds of the log export volume (Beuter and
authors, 1976; Gedney and coauthors, 1975; Oswald, 1978). After 1990,
a increasing proportion of harvest on industry lands will be second-
:owth timber. We expect that the decline in harvest and the change in
ie character of the timber will work against export sales.

Table 3. Imports and Exports of Softwood Logs for the United States in
Selected Years, 1960-2030

(billion board-feet, International 1/4-inch log scale)

Year	Imports	Exports
1960	a	0.3
1965	a	1.5
1970	0.1	3.4
1971	0.1	2.8
1972	0.1	3.8
1973	0.1	3.9
1974	0.1	3.2
1975	0.1	3.3
1976	0.1	4.0
1977	0.2	3.9
1990	0.1	3.8
2000	0.1	3.3
2010	0.1	3.0
2020	0.1	2.8
2030	0.1	2.5

Source: Data for 1960-77 are from Phelps, R.B. 1977. The Demand and
Price Situation for Forest Products, 1976-77, U.S. Department of Agricul-
ture Miscellaneous Publication 1357 (Washington, D.C., U.S. Department of
Agriculture).

[a]Less than 50 million board-feet.

Most of the U.S. log volume exported to Japan is processed into lumber for use in Japan's housing industry. Available information (Ueda and Darr, 1980) points to relatively stable or possibly even declining numbers of housing starts over the next two decades, depending on assumptions made about the rate of replacement of Japanese housing stock. The possibility of stable or declining housing demand in Japan contrasts sharply with the rapid increase in housing starts during the 1960s and early 1970s. We expect the future demand situation to have a downward effect on U.S. log export volumes.

Even a cursory examination (Gallagher, 1979) of Japanese timber inventory, growth, and harvest data suggests significant potential for increased harvesting of timber in Japan. An increase in domestic harvest would be in sharp contrast to the pattern of the 1960s and 1970s, when harvest was generally declining. There is much uncertainty, however, about when domestic harvesting might begin to increase. Exchange rates, government forestry policies, trade policies—a mix of many factors—will determine the timing and extent of realization of the Japanese harvest potential. We assumed that some increase in the Japanese harvest would occur, decreasing the demand for U.S. softwood logs.

We assumed that other sources of softwood logs for Japan—primarily the Soviet Union and New Zealand—would continue future shipments at near current levels.

The long-range prospects for shipments of tropical hardwood to Japan are uncertain. The actions of a successful hardwood cartel composed of Indonesia, Malaysia, and the Philippines might cause Japan to substitute softwood for hardwood plywood in some end uses. This could affect the demand for imported softwood logs. There is no history of significant use of softwood plywood in Japan, however, and to date, consumption of tropical hardwoods by the Japanese has not been sensitive to price changes.

Projections of domestic U.S. demands indicate continuing strong markets for softwood construction materials. This would help the competitiveness of domestic processors on the West Coast and thereby dampen log exports.

In summary of the projections for softwood log exports, most of the factors potentially affecting U.S. sales point in a downward direction in our judgment, and we made our projections accordingly.

Pulpwood

As in the case of softwood logs, we expect U.S. exports of pulpwood--roundwood and chips--to peak about 1990 and then to gradually decline throughout the projection period (table 4). Until the mid-1960s, U.S. exports of pulpwood consisted mainly of shipments of roundwood to Canada. In 1965 Japan began buying chips from the West Coast of the United States. The volume of these shipments increased rapidly and now amount to over 3 million tons, valued at about $150 million per year. Japan takes about 95 percent of total U.S. exports of chips. The export of chips from the South to Scandinavia has not lived up to initial expectations.

The projections of domestic demand for paper and board products indicate strong market growth---to the point that significant volumes of chips may not be available for export. In part, these projections led us to expect a downturn in export shipments. Also, the expected decline in harvest on industry lands on the West Coast will probably reduce the pro-

Table 4. Imports and Exports of Pulpwood for the United States in Select-
          ed years, 1960-2030

(million cords)

Year	Imports	Exports
1960	1.5	0.2
1965	1.3	.2
1970	1.1	1.8
1971	1.2	1.5
1972	1.0	1.9
1973	1.2	2.6
1974	.9	3.0
1975	.7	3.0
1976	1.0	3.6
1977	1.2	3.7
1990	1.3	4.8
2000	1.3	4.0
2010	1.3	3.5
2020	1.3	3.0
2030	1.3	2.8

Source: Data for 1960-77 are from Phelps, R.B. 1977. The Demand and
Price Situation for Forest Products, 1976-77, U.S. Department of Agricul-
ture Miscellaneous Publication 1357 (Washington, D.C., U.S. Department of
Agriculture).

duction of mill residues since lumber output will decline as harvest is reduced. In addition, Japan now has many sources of chip supplies, such as Australia, South Africa, South America, and Canada. This is in contrast to the 1960s when the United States was almost the only supplier. The availability of chips from several sources will dampen growth in Japanese demand for U.S. chips. An increase in domestic harvest in Japan would also dampen demand for imported chips.

There is probably a limit to development of pulp mills in Japan. The lack of availability of sites and environmental problems associated with concentrated development of pulp mills will probably limit growth in pulp production. We assumed that these types of constraints on pulp production in Japan would lead to an expansion of the markets for pulp, paper, and paperboard--products to be discussed later.

To summarize our rationale for pulpwood exports, most supply and demand factors seem to point in a downward direction for export volume. These factors are reflected in our projections. Pulpwood imports are expected to remain near current levels throughout the projection period.

Softwood Lumber

In the assessment projections, U.S. consumption of softwood lumber generally follows the track of housing starts (table 5). Housing starts are expected to peak at 2.6 million units in 1990 and then decline to 2 million in 2030.

Imports of softwood lumber into the United States, almost all from Canada, follow a similar pattern--peaking in 1990 and then decreasing. Some increase in U.S. production of softwood lumber is also expected.

Table 5. Production, Imports, Exports, and Consumption of Softwood Lumber and Number of Housing Starts for the United States, in Selected Years, 1960-2030

(billion board-feet)

Year	Production	Imports	Exports	Consumption	Housing starts (millions)
1960	26.7	3.6	0.7	29.6	1.2
1965	29.3	4.9	0.8	33.4	1.5
1970	27.5	5.8	1.2	32.1	1.5
1971	30.0	7.2	0.9	36.3	2.1
1972	31.0	9.0	1.2	38.8	2.4
1973	31.6	9.0	1.8	38.8	2.1
1974	27.7	6.8	1.6	32.9	1.4
1975	26.7	5.7	1.4	31.0	1.2
1975	30.8	8.0	1.6	37.2	1.5
1977	31.1	10.4	1.4	40.1	2.0
1990	36.2	13.5	1.9	47.8	2.6
2000	37.4	13.0	1.8	48.6	2.2
2010	41.1	13.0	1.7	52.4	2.3
2020	41.1	12.5	1.6	52.0	2.3
2030	40.9	12.0	1.6	51.3	2.0

Note: All projections for softwood lumber are for medium, base-level series (U.S. Forest Service, 1979. An Assessment of the Forest and Range-land Situation in the United States, review draft (Washington, D.C.).

Source: Data for 1960-77 are from Phelps, R.B. 1977. The Demand and Price Situation for Forest Products, 1976-77, U.S. Department of Agriculture Miscellaneous Publication 1357 (Washington, D.C., U.S. Department of Agriculture).

Imports of softwood lumber from Canada will depend in part on the competitiveness of U.S. producers, offshore and Canadian demands for Can. dian lumber, U.S. demands, and the availability of the timber resource i Canada.

Much of the expansion of U.S. imports from Canada in the 1960s and 1970s was made possible by increased production in British Columbia. About one-half the physical allowable annual cut is currently being used in Canada. There is general agreement on the potential for significant expansions of output of all forest products. There is little agreement, however, on just how much additional production might be supported. Reports by F. L. C. Reed and Associates (1974, 1977) are probably the most comprehensive assessment of the condition of the Canadian timber resourc These reports give an estimate of the maximum physical allowable annual cut and the economically accessible allowable annual cut. Data derived from these reports suggests that the Canadian resource could support sof wood lumber production of about 19 billion board-feet if the economicall accessible allowable cut is fully utilized and of about 21 billion board feet if the physical allowable annual cut is fully committed. This compares with current production of 17 billion board-feet. Arguments could support higher or lower maximums for lumber production potential. The point is that the Canadian resource will not, over the coming decades, support the kinds of expansion that occurred in the 1960s and 1970s. Th view is reflected in our projections of softwood lumber imports.

Some growth in demand in Canada is expected, and there will probabl be some growth in Canadian offshore shipments to Japan and western Europ

(Aird and Ottens, 1979). These competing demands will dampen growth in sales to the United States.

Our projections of exports of softwood lumber from the United States follow a pattern similar to softwood logs; that is, increasing somewhat to 1990 but then tapering off throughout the projection period. In part, these projections are based on the expected decline in harvest on industry lands on the West Coast. About two-thirds of U.S. exports originate on the West Coast. In addition, softwood log exports from lands granted to Native corporations in southeast Alaska will probably become much more important over the next decade or so. These logs may substitute for lumber, especially Sitka spruce cants, currently exported from southeast Alaska to Canada. Current log export restrictions prohibit log exports from federal lands, but these restrictions do not apply to land granted to Native corporations.

Currently, Canada takes about 28 percent of U.S. softwood lumber exports; Japan, 30 percent; western Europe, 20 percent; Australia, 11 percent; and other countries, the remaining 11 percent. These shipments are heavy on clears, large timbers, and other specialty items not generally characterized as fast-growth markets. Efforts to promote the sale of dimension lumber have met with only limited success in Japan and western Europe. We assumed that any future success of these efforts would show up more in additional drains on the Canadian resource rather than on the U.S. resource.

Throughout the projection period, imports from Canada would amount to 30 to 35 percent of U.S. consumption of softwood lumber. Exports from the United States would amount to 5 percent of production.

Softwood Plywood

Consumption of softwood plywood in the United States is projected to increase throughout the projection period (table 6).

Tariff and nontariff trade barriers are expected to continue to limi U.S.-Canadian trade in softwood plywood. Thus, U.S. imports are expected to be negligible throughout the projection period.

Exports of softwood plywood are expected to peak at 900 million square feet in 1990 and to decline gradually through the remainder of the projection period. The rationale for the export pattern is based in part on the expected timber supply on the West Coast. About 80 percent of U.S exports originate in the Pacific Northwest. We assumed that fiber-based building boards increasingly will be substituted for softwood plywood in Europe, thereby dampening any increase in demands for imports. As discussed previously, the tropical hardwood situation and its effects on the hardwood plywood industry in Japan may lead to an expanded Japanese market for imported softwood plywood. We have not, however, made any allowance for expanded shipments to Japan in our projections.

Throughout the projection period, U.S. exports of softwood plywood are expected to account for less than 5 percent of production.

Hardwood Lumber

Of the solid timber products—logs, lumber, and plywood—production of and trade in hardwood lumber are probably the most difficult to projec Production of hardwood lumber in the United States has been relatively stable since 1950. Information on the hardwood timber resource situation however, indicates a continuing buildup of hardwood inventories. For the

Table 6. U.S. Production, Imports, Exports and Consumption of Softwood
Plywood for the United States in Selected Years, 1960-2030

(billion square feet, 3/8-inch basis)

Year	Production	Imports[a]	Exports	Consumption
1960	7.8		a	7.8
1965	12.4		a	12.4
1970	14.3		0.1	14.2
1971	16.6		0.1	16.5
1972	18.3		0.2	18.1
1973	18.3		0.5	17.8
1974	15.9		0.5	15.4
1975	16.1		0.8	15.3
1976	18.4		0.7	17.7
1977	19.4		0.3	19.1
1990	24.6		0.9	23.7
2000	25.1		0.8	24.3
2010	27.3		0.7	26.6
2020	27.8		0.6	27.2
2030	27.9		0.6	27.3

Note: All projections for softwood plywood are for medium, base-level series, (U.S. Forest Service, 1979. An Assessment of the Forest and Rangeland Situation in the United States, review draft (Washington, D.C.).

Source: Data for 1960-77 are from Phelps, R. B. 1977. The Demand and Price Situation for Forest Products, 1976-77, U.S. Department of Agriculture Miscellaneous Publication 1357 (Washington, D.C., U.S. Department of Agriculture).

[a]Less than 50 million square feet.

most part, the buildup of inventory is in lower-quality hardwoods. The projections indicate rapidly expanding domestic demands for pallets and other end products that can use hardwoods and do not have lumber quality as a key criterion. Projections for domestic lumber production assume that the resource and demand situations will result in significant expansion of output throughout the projection period (table 7).

Imports of hardwood lumber are expected to increase gradually throughout the projection period. In part expanded imports reflect expected increases in demand and in part expected increases in the availability of tropical hardwood lumber. Most countries with significant inventories of tropical hardwoods are taking steps to increase the export of processed end products at the expense of logs. The United States is not a major importer of tropical hardwood logs, but increased availability of lumber on world markets will probably be reflected in increased U.S. imports.

We expect some expansion of U.S. hardwood lumber exports, primarily to western Europe. For the most part, these shipments will probably be higher-quality hardwoods, such as oak.

Even with some expansion of trade, foreign markets will continue to be relatively minor factors in U.S. markets for hardwood lumber. For example, annual exports never exceed 5 percent of production during the projection period and imports never exceed 7 percent of consumption.

Hardwood Plywood

Hardwood plywood production in the United States is expected to increase gradually throughout the projection period (table 8). The relatively slow growth in imports reflects our expectation that depletion of

Table 7. Production, Imports, Exports, and Consumption of Hardwood Lumber, for the United States in Selected Years, 1960-2030

(billion board-feet)

Year	Production	Imports	Exports	Consumption
1960	6.3	0.3	0.3	6.4
1965	7.5	0.3	0.1	7.7
1970	7.1	0.3	0.1	7.3
1971	6.9	0.4	0.2	7.1
1972	6.8	0.4	0.3	7.0
1973	7.0	0.5	0.2	7.3
1974	6.9	0.5	0.2	7.2
1975	5.9	0.2	0.2	5.9
1976	6.4	0.3	0.2	6.5
1977	6.2	0.3	0.2	6.3
1990	10.0	0.4	0.2	10.2
2000	11.3	0.6	0.3	11.6
2010	12.9	0.8	0.4	13.3
2020	14.8	0.9	0.5	15.2
2030	16.3	1.0	0.5	16.8

Note: All projections for hardwood lumber are for medium base-level
series, (U.S. Forest Service, 1979. An Assessment of the Forest and
Rangeland Situation in the United States, review draft (Washington, D.C.).

Source: Data for 1960-77 are from Phelps, R. B. 1977. The Demand
and Price Situation for Forest Products, 1976-77, U.S. Department of Agriculture Miscellaneous Publication 1357 (Washington, D.C., U.S. Department
of Agriculture).

Table 8. Production, Imports, Exports, and Consumption of Hardwood
Plywood for the United States, in Selected Years, 1960-2030

(billion square feet, 3/8-inch basis)

Year	Production	Imports	Exports	Consumption
1960	1.1	0.7	a	1.8
1965	2.0	1.0	a	3.1
1970	1.8	2.0	0.1	3.8
1971	1.9	2.5	a	4.5
1972	2.1	3.2	a	5.2
1973	1.9	2.5	a	4.4
1974	1.6	1.7	0.1	3.1
1975	1.2	1.9	0.1	3.1
1976	1.4	2.4	0.1	.37
1977	1.5	2.3	0.1	3.6
1990	1.8	3.5	a	5.3
2000	2.1	3.7	a	5.8
2010	2.2	4.0	a	6.2
2020	2.6	4.0	a	6.6
2030	3.0	3.8	a	6.8

Note: All projections for hardwood plywood are for medium, base-level series.

Source: Data for 1960-77 are from Phelps, R. B. 1977. The Demand and Price Situation for Forest Products, 1976-77, U.S. Department of Agriculture Miscellaneous Publication 1357 (Washington, D.C., U.S. Department of Agriculture).

[a]Less than 50 million square feet.

tropical hardwood inventories will become an important factor in world markets by the year 2000. There seems to be a consensus that inventories are being depleted, but there is uncertainty about the timing of downfalls in production for individual countries (Pringle, 1976). Current efforts at phasing out log exports in Indonesia, Malaysia, and the Philippines may affect the country from which U.S. hardwood imports originate but will probably have little other effect on U.S. supplies.

Exports of hardwood plywood are expected to be negligible throughout the projection period. Imports will account for about two-thirds of domestic consumption throughout the projection period.

Wood Pulp

Production of wood pulp in the United States nearly doubled between 1960 and 1977. Substantial increases in output are expected over the coming decades, but the rate of increase is projected to be slower than in recent decades (table 9).

Both imports and exports of wood pulp are expected to grow gradually throughout the projection period. We expect Canada to continue as the primary source of imports. There is potential for significant expansion of pulp output in Ontario, Quebec, the Prairie Provinces, and British Columbia. As for softwood lumber, however, there is no consensus about the ultimate potential for expansion of the Canadian pulp industry. The continuing expansion of the U.S. market, however, and the potential of close corporate ties between Canadian and U.S. producers should cause continued expansion of U.S. imports.

Table 9. Production, Imports, Exports, and Consumption of Wood Pulp for
          the United States, Selected Years, 1960-2030

(million tons)

Year	Production	Imports	Exports	Consumption
1960	25.3	2.4	1.1	26.6
1965	34.0	3.1	1.4	35.7
1970	43.5	3.5	3.1	44.0
1971	43.9	3.5	2.2	45.2
1972	46.8	3.7	2.3	48.2
1973	48.3	4.0	2.3	50.0
1974	48.3	4.1	2.8	49.7
1975	43.1	3.1	2.6	43.6
1976	38.8	3.7	2.5	50.0
1977	50.0	3.9	2.6	51.2
1990	71.6	5.7	3.2	73.6
2000	85.9	6.1	3.7	88.3
2010	100.2	7.0	4.0	103.2
2020	115.1	7.6	4.2	118.5
2030	128.9	7.9	43.	132.5

Note: All projections for wood pulp are for medium, base-level
series.

Source: Data for 1960-77 are taken from Phelps, R. B. 1977. The
Demand and Price Situation for Forest Products, 1976-77, U.S. Depart-
ment of Agriculture Miscellaneous Publication 1357 (Washington, D.C., U.S.
Department of Agriculture).

Our expectation of increased exports of wood pulp is based on the assumption of continuing growth in demand in Japan and western Europe. There is little doubt that demand for pulp will grow in these two areas (Food and Agricultural Organization, Economic Commission for Europe 1976; Japan Ministry of Agriculture and Forestry, 1973), but there are questions about what rates of growth to expect. Japan is still struggling under the escalation of energy costs since the 1973-74 period, and future rates of economic growth are not likely to reach the double-digit rates of the 1960s and early 1970s. Similarly in western Europe, the future for economic growth seems uncertain.

Additional market factors considered in our projections were the timber supply situations in Scandinavia and the tropical areas. Producers in Sweden, a traditional major supplier of pulp to western Europe, are restricted by production ceilings imposed by timber supply constraints. These producers are also taking steps to upgrade their product mix for exports--more finished paper and board products and less pulp, for example. These market charactierstics point to increased demands for U.S. pulp.

On the other hand, Brazil and other tropical countries are on their way to becoming significant producers of market pulp that may end up in Japanese and western European markets over the coming decades. In addition it should be kept in mind that about one-third of U.S. export volume is currently composed of dissolving and special alpha, grades having a declining market prospect.

Imports of wood pulp are expected to continue to account for 7 percent to 8 percent of U.S. consumption and exports 4 percent to 5 percent of production throughout the projection period.

Paper and Paperboard

Production of paper and board in the United States is expected to triple by 2030, as compared with current output (table 10). The projections reflect the assumption of increased recycling of paper and board, and this accounts in part for more rapid expansion of paper and board out put compared with wood pulp. Both imports and exports of paper and board are expected to increase gradually throughout the projection period.

It is expected that most of the imports will continue to be dominated by newsprint. Producers of newsprint in the United States are also expected to increase their share of the U.S. market. The market is expected to grow, however, leading to increased imports despite increased domestic production. As discussed previously, the timber supply situation in Canada may limit Canadian production, especially after 2000.

Currently, U.S. exports of paper and board are primarily to western Europe (29 percent), Central and South America (26 percent), and Canada (22 percent). Japan accounts for less than 5 percent of sales. We expect growth in demand for U.S. output in both Japan and western Europe. In part, this reflects the likelihood of joint ventures between U.S. producers and Japanese and European investors. Economic growth in Japan and western Europe will lead to increased demand in the face of increasingly tight supplies of wood fiber. U.S. consumption is expected to triple ove the next fifty years, however, and growth in the domestic market will probably inhibit the development of export sales.

Exporters in the United States are likely to continue to face opposition from domestic producers in both Japan and western Europe. The recently completed Multilateral Trade Negotiations (MTN) provide for some

Table 10. Production, Imports, Exports and Consumption of Paper and
          Board Products for the United States, for Selected Years,
          1960-2030

(million tons)

Year	Production	Imports	Exports	Consumption
1960	34.4	5.3	0.3	39.3
1965	44.1	6.8	1.8	49.1
1970	53.5	7.3	2.8	57.9
1971	55.1	7.6	3.0	59.7
1972	59.5	8.0	2.9	64.5
1973	61.3	8.4	2.8	66.9
1974	59.9	8.4	3.5	64.8
1975	52.5	6.3	3.2	55.7
1976	60.0	7.2	3.5	63.8
1977	60.7	7.6	3.3	65.1
1990	95.5	9.3	4.5	100.3
2000	118.0	10.3	4.9	123.4
2010	141.9	11.2	5.3	147.8
2020	165.5	12.0	5.7	171.8
2030	187.9	12.7	6.0	194.6

Note: All projections for paper and board products are for medium,
base-level series.

Source: Data for 1960-77 are from Phelps, R. B. 1977. The Demand
and Price Situation for Forest Products, 1976-77, U.S. Department of Agriculture Miscellaneous Publication 1357 (Washington, D.C., U.S. Department
of Agriculture).

reductions in tariff barriers on paper and board products in these two areas. A period of slow economic growth, however, coupled with fluctuati business cycles would make foreign producers fight U.S. sales through trade restrictions.

We have made only limited allowance for expansion of U.S. exports to developing areas. Although per capita consumption in these areas is low compared with the industrialized countries, just about every developing country with a forest resource seems to be investigating the possibility of constructing pulp mills. In most cases, substitution of domestic production for imports is the primary reason for considering construction of pulp or paper mills, or both. This type of development will limit U.S. sales.

According to our projections, imports of paper and board as a percentage of U.S. consumption would decline from about 12 percent in the late 1970s to about 8 percent in 2030. Exports as a percentage of production would decline from 5 percent to about 3 percent.

Discussion of Projections

In general, our projections indicate a continuation of current U.S. trade patterns for timber products. Imports will continue to be dominated by softwood lumber, pulp, newsprint from Canada, and hardwood plywood from Southeast Asia. Exports will consist primarily of softwood logs, chips, softwood lumber, pulp, and paper and board, with Japan and western Europe as the major customers. Most people would probably not object to our assessment of trade prospects in these terms. Areas of disagreement seem to center on the amount of imports and exports. Are our projections too high or too low?

Our projections have been available for comment for some time. In general, there seems to be little disagreement with our projections of imports. There appears to be a wide range of opinion about potential for U.S. exports, however. Individuals and organizations who feel that our projections are too conservative generally base their arguments on the following rationale.

The United States, Canada, Scandinavia, and the Soviet Union are the primary sources of softwood timber products entering international trade. Available information indicates a tightening of supplies in Canada, Scandinavia, and the Soviet Union. According to this rationale, a tightening of supplies from competitors will open up significant opportunities for U.S. producers to export products to Japan and western Europe, especially pulp, paper, and board products. The United States has some of the most productive forestland in the world and could produce enough timber to supply both domestic and export markets if an appropriate mix of trade and public timber supply policies could be implemented. Carried to the extreme, this rationale becomes an advocacy position for the United States to become the "wood basket" of the world. Is this a reasonble rationale? If it is reasonable, why are our projections not more optimistic?

Our projections were influenced by historical trends in U.S. trade, projections of domestic U.S. markets, and our judgment of how world markets for timber products will interact over the coming decades. We have assumed no significant changes in U.S. trade and timber supply policies that would be aimed specifically at imports and exports of timber products.

Thus, we have made no allowance for the potential success of policies designed to increase exports.

The historical data on trends in U.S. exports and imports reflect internactions of markets, comparative advantage, tariff and nontariff trade barriers--all factors that influence trade. The period of the 1950s, 1960s, and early 1970s was one of unparalleled economic growth and world trade. The economies of Japan and western Europe were recovering from war, the U.S. economy was experiencing a period of sustained economic growth of record length, MTN reduced tariff barriers in the industrialized countries, cheap energy was the rule of the day, and the developing countries were not yet challenging the industrialized countries regarding the distribution of gains from trade in natural resources. Exports of timber products from the United States in the 1960s and early 1970s responded to these types of stimuli, primarily in the form of logs, lumber, chips, and pulp from the Pacific Northwest to Japan; pulp and lumber from Alaska to Japan; and pulp, paper, and paperboard from southern United States to western Europe. Despite the growth in worldwide markets for timber products during the past two decades, the U.S. industry is still oriented to domestic markets. Exports of most timber products amount to less than 10 percent of production. In comparison, about one-third of U.S. agricultural output is exported.

Looking ahead, is there anything that is likely to change the apparent domestic orientation of the U.S. timber industry? Certainly, some members of U.S. industry are attempting to build support for policies designed to expand U.S. exports (National Forest Products Association, 1979). These efforts may pay off, but the environment for their operation will be much different from that of the 1960s and early 1970s. The U.S. economy may have an indigeneous inflation rate as high as 9 percent; pronounced business cycles are once again characteristic of the U.S. economy; the Japanese and western European economies have matured; and nontariff trade barriers have proliferated during the 1970s and will probably continue to be major factors in world trade despite the best intentions of participants in the recent Multilateral Trade Negotiations. In addition, the developing countries will continue to press for a bigger piece of the pie, and energy costs will go up, perhaps catastrophically, if events in the Middle East cannot be controlled. The only characteristic of the U.S. market for timber products over coming decades that will probably be similar to the postwar period is that domestic markets will continue to be strong, especially for paper and board. Significant expansion of exports in the face of this type of environment will be difficult at best.

In conclusion, I would like to raise two questions; Are there better ways to do these types of projections of long-term trade? Does it make sense to try to project what trade might be fifty years into the future? The Forest Service will probably be making these long-term assessments every five years, so the question of methodology for projecting trade is not one that will go away.

REFERENCES

Aird, K. L., and J. Ottens. 1979. The Outlook for Timber Utilization in Canada to the Year 2000 (Ottawa, Ontario, Canadian Forestry Service, Policy, Development and Analysis Branch).

Beuter, John H., K. H. Johnson, and H. C. Scheurman. 1976. Timber for Oregon's Tomorrow: An Analysis of Reasonably Possible Occurrences, Research Bulletin 19 (Corvallis, Forest Research Laboratory, School of Forestry, Oregon State University).

Food and Agricultural Organization, Economic Commission for Europe. 1976. "European Timber Trends and Prospects, 1950 to 2000," Timber Bulletin Europe vol. 29 (Suppl. 3).

Gallagher, Paul. 1979. "An Analysis of the Softwood Log Trade Between the United States and Japan," draft on file at Pacific Northwest Forest and Range Experiment Station, Portland Oregon.

Gedney, D. R., D. D. Oswald, and R. D. Fight. 1975. Two Projections of Timber Supply in the Pacific Coast States. U.S. Forest Service Resources Bulletin PNW-60 (Portland, Oregon, Pacific Northwest Forest and Range Experiment Station).

Japan Ministry of Agriculture and Forestry. 1973. Basic Plan Regarding Forest Resource and Long Range Prospect Regarding Demand and Supply of Important Forest Products. Approved at the Cabinet Council, February 16, 1973 (Tokyo).

National Forest Products Association. 1979. International Trade Report 15 (Washington, D.C., International Trade Committee).

Oswald, Daniel D. 1978. Prospects for Sawtimber Output in California's North Coast, 1975-2000. U.S. Forest Service Resource Bulletin PNW-74 (Portland, Oregon, Pacific Northwest Forest and Range Experiment Station).

Phelps, R. B. 1977. The Demand and Price Situation for Forest Products, 1976-77, U.S. Department of Agriculture Miscellaneous Publication 1357. (Washington, D.C., U.S. Department of Agriculture).

Pringle, S. L. 1976. "Tropical Moist Forests in World Demand, Supply and Trade," Unasylva vol. 28 (112-113).

. L. C. Reed and Associates. 1974. Canada's Reserve Timber Supply, prepared for the Department of Industry, Trade and Commerce (Ottawa, Ontario).

____. 1977. Forest Management in Canada, vol. 1, prepared for the Forest Management Institute of the Canadian Forestry Service (Ottawa, Ontario).

eda, Michikiki, and David R. Darr. 1980. "The Outlook for Housing in Japan to the Year 2000," review draft on file at Pacific Northwest Forest and Range Experiment Station, Portland, Oregon.

.S. Department of Commerce. 1978a. U.S. Exports: Commodity by Country, FT-410 (Washington, D.C.).

____. 1978b. U.S. Imports: Commodity by Country, FT-135 (Washington, D.C.).

.S. Forest Service. 1979. An Assessment of the Forest and Rangeland Situation in the United States, review draft (Washington, D.C.).

Discussion by John Zivnuska

David Darr has effectively summarized a complex Forest Service analysis of international trade. His paper presents a projection of trade in
forest products developed within the framework of a particular model of
domestic consumption with the United States, but it does not examine that
model. To comment on the Forest Service projection, however, I must dire
as much attention to the framework within which the trade analysis was developed as to the analysis itself.

This analysis was prepared as a part of the 1979 Assessment under the
Resources Planning Act, with the data summarized pertaining to what is
termed the medium-level projection. The projections of consumption, production, and trade are developed in the framework of the familiar "gap"
model which has been used in Forest Service studies at least since World
War II. This gap model is used for analytical purposes, and the projections developed under it clearly do not reflect the actual developments
which are considered to be likely.

Under the gap model, fixed assumptions are made as to future price
trends for the major groups of forest products. The quantities which wou
be demanded at such future prices are then projected. The quantities
which would be supplied under comparable assumptions as to price are also
projected. However, as the Forest Service tells us, "the rate of increas

in product prices during the projections period is likely to be higher
than assumed" (U.S. Forest Service, 1979). As is to be expected, this use
of unrealistically low price trend assumptions results in projections of
demands which are higher than the projections of supply, thus showing a
gap between consumption and production.

Actually, such a gap cannot and will not occur. Again, as the Forest
Service tells us, prices will rise substantially above the assumed levels,
the quantities demanded will be reduced, the quantities supplied will be
increased, and a condition approximating market equilibrium will be realized. In the 1979 Assessment, the Forest Service for the first time provides some projections as to quantities supplied and demanded at equilibrium price levels in addition to the much more detailed estimates under
the gap model. For example, the quantity of softwood timber demanded from
the forests of the United States in the year 2000 is projected at 12.2
billion cubic feet under equilibrium prices as compared to 13.8 billion
cubic feet under the unrealistically low price assumptions of the gap
model--a reduction in projected domestic softwood consumption of some 1.6
billion cubic feet, or 12 percent.

Unfortunately, the findings of the equilibrium analysis are not presented in sufficient detail to enable any overall estimation of the effects
of using equilibrium prices rather than unrealistic price assumptions on
the Forest Service projections of the domestic consumption, domestic production, imports, and exports of the major forest products. It seems evi-
dent, however, that the assignment given the agency's Foreign Trade Analysis Work Unit in Portland was to develop projections of international
trade in forest products within the framework of domestic consumption
levels as projected in the gap model rather than in the framework of the

appreciably lower domestic consumption levels actually to be expected as
indicated by the equilibrium price analysis.

Thus, the Forest Service trade projections appear to have been appreciably influenced by being tied to domestic consumption projections which
are higher than will be realized under equilibrium prices. If the dis-
torting effects of these hypothetical consumption levels could somehow be
removed from the Forest Service international trade projections, the outlook shown would differ significantly from that which is now displayed.

In this regard, I must raise the question as to the exact meaning of
the production projections which are shown in tables 5 through 10 of Darr's
paper. As nearly as I can determine, these are not projections of the
amounts which will be produced in 1990 and·beyond, but instead are projections of the amounts which would have to be produced to support the projected consumption levels as adjusted for the projections of imports and
exports which are displayed. Upon further reading of the 1979 Assessment,
one discovers that the amounts of timber supplied under the gap model
price assumptions will not be sufficient to support such production levels.
Thus the gap between consumption and production appears, and it is shown
that the whole balance among production, imports, exports, and consumption
which is displayed is not believed to be attainable.

If all this sounds confusing, I can assure you that it is. Such con-
fusion is simply part of the price which must be paid for the Forest Service's predilection for basing its timber assessment on a gap model of
that which will not happen and then presenting its findings in terminology
which is not consistent with normal economic usage. In any case, for
those concerned with international trade in forest products the immediate
question is the extent to which the tie to unrealistic domestic consumption

and price projections has influenced the projections of imports and exports developed by Darr's group. I cannot attempt to answer this comprehensively, but a quick look at some major product groups may serve to illustrate the nature of the problem and show something of its possible magnitude.

The gap model consumption projections to which the trade projections are tied assume a rate of change in relative prices of 0.7 percent per year for softwood lumber and no relative change for softwood plywood. However, the Forest Service reports that the equilibrium price analysis "shows softwood lumber and plywood increasing at an annual rate of 1.8 and 1.4 percent respectively." The differences are tremendous. At the 1.8 percent trend, the real price of lumber in 2000 would be 51 percent above the base year, while the gap model consumption projection allows for only a 17 percent increase. With allowance for the effects of this equilibrium price path on the quantity demanded, softwood lumber consumption in 2000 would be projected at slightly more than 40 billion board-feet rather than more than 48 billion board-feet as now shown. This is about the 1977 consumption level, which indicates that even under the assumption of no increase in U.S. lumber production, the level of imports from Canada will not increase above recent peak year levels. To the degree that U.S. production rises in response to increasing prices, some modest decline from the peak 10 billion board-feet import level is possible. While this leaves the United States as a major net importer of softwood lumber, it is a quite different outlook than the increase in imports of nearly 30 percent projected under the gap model consumption level for 2000. (In this regard, it should be remembered that the projected U.S. production level of 37.4

billion board-feet shown for the year 2000 in Darr's table 5 is apparently
considered not to be attainable. If it were to be realized, a major reduction in the level of softwood lumber imports from Canada would be in prospect.)

I must add that this outlook for no further increases and possibly
some modest decline in the trend level of softwood lumber imports from
Canada appears to be consistent with the forest situation in Canada. While
the Canadians appear to have some limited opportunity to expand cutting
levels in the forests which are now or will soon be accessible, as well as
some additional possibilities in forests which may become accessible in the
next century, any such increases will not be in the species, sizes, and
grades which are the basis of the Canadian export trade in lumber to the
United States. In addition, Canadian producers face expanding opportunities in offshore markets and have shown a definite interest in reducing
their present high degree of dependence on U.S. markets.

In the general group of wood pulp, paper, and paperboard, the present
import dependency of the United States is based entirely on imports of
newsprint paper and wood pulp from Canada. The gap model projections of
the Forest Service and the related trade projections assume no increase in
the relative price of these products. Actually, a definite upward trend
in these relative prices is necessary, both to attract the investment
capital needed to expand capacity and to cover the upward trend in the
costs of pulpwood delivered to the mill which must be expected. The Forest
Service states that some upward trend in equilibrium prices is expected,
but does not state the amount of this increase. Thus it is not possible
directly to estimate the effects of allowing for equilibrium prices in
changing the gap model projections of the Forest Service. On the basis of

:her evidence I believe that the equilibrium price level consumption of
per and board in 2000 is likely to be somewhat less than 110 million tons,
ther than the 118 million tons shown by the Forest Service under the no-
al-price-increase assumption. On this basis, the net import dependency
 the United States would drop to about 2 million tons by 2000, rather
an expanding to nearly 5.5 million tons as shown by the Forest Service.

The import dependency of the United States in paper and board is
sed entirely on imports of newsprint paper from Canada. Some modest in-
ease in such imports seems likely. However, in all other grades of paper
d board combined the United States is already a net exporter to the ex-
nt of 2.5 million tons. Such net exports could be increased to nearly 6
llion tons by 2000. The most favorable opportunities for export expan-
on appear to be in linerboard and folding boxboard. In wood pulp, the
ited States has been a net importer of about 1.5 million tons annually
 recent years. With adjustments in production and consumption in res-
nse to rising prices, the nation could become a net exporter to nearly
e same degree by 2000.

This is far from suggesting that the southern United States is about
 become the woodbox of the world. I do suggest, however, that the south-
n United States has advantages in the production of some large-volume
od products and that this can be important to the nation's balance of
ade.

In the case of softwood log exports, I again differ from the Forest
rvice projections, but on very different grounds. Such exports are domi-
ntly from the Pacific Northwest to Japan. On the supply side, the Forest
rvice projections of timber harvest for the Pacific Coast region assume
at the cutting level on the federal lands will be sustained at recent

levels through 2000 and that an increase in cut from small ownership in response to rapidly increasing stumpage prices will tend to offset the ex pected decline in cut from industrial lands. In this I believe the agenc has underestimated the impact of various policy, budgetary, and multiple-use constraints in reducing the federal cut and also the combined effect environmental constraints and owner objectives in holding down any increa in harvest from smallholdings. Thus I believe the supply situation will much tighter than the Forest Service projects. I suggest that market forces reflecting the effort to maintain existing mill capacity and to su tain the current level of lumber production in the face of a declining ti ber harvest will lead to a much sharper reduction in log exports by the e of the century than the gradual decline projected by the Forest Service.

All my comments have been limited to the period between now and 2000 Darr raises the question of projections extending for up to fifty years, as is required of the Forest Service under the Resources Planning Act. I believe present projection methods are wholly unsuitable for projections such length. Present methods involve the pyramiding of assumption upon assumption. Within even twenty years, the whole structure becomes so un-stable as to be unable to support further extension. In addition, in attempting to move into the next century the forest products analyst is left wholly without guidelines as to probable levels of population, econo mic activity, and forest conditions among the many nations and regions which are involved in the increasingly intricate network of international trade in forest products. There is simply no general economic framework for the world within which to develop forest products trade projections e tending to 2030. Even major uncertainties as to new sources of timber emerge by the next century. For example, I believe there is good reason to anticipate only limited impact on the major markets of interest to North America during the next twenty years from the type of fast-growing plantations which Roger A. Sedjo has discussed. In the following decades, however, such timber might become a factor in international trade.

Darr also raises the question of the possible effects of the present economic difficulties of the United States on the ability of domestic producers to capture export markets. I must agree with him that nothing in the pattern of recent years is consistent with the overall economic setting projected for the 1979 Assessment. The projection of an economy by the year 2000 characterized by a 75 percent increase in real per capita disposable personal income, along with a substantial increase in leisure time, seems more and more to be an exercise in fantasy for a nation in which the annual growth in real GNP per employed worker averaged only 0.1 percent over the period from 1973 through 1979. With the lowest level of per capita savings and capital investment of any major industrial nation, with per capita productivity now actually declining, and with an inflation rate which continues to accelerate, and with a huge unfavorable balance in international trade, the United States is traveling the road to disaster, not to a more abundant future.

But in terms of forest products, I see opportunity and need in this situation, not just problems. One must believe that this nation will finally recognize the root causes of its difficulties and set about to solve them. One element among the major structural shifts which are needed in the economy is surely the expansion of the export trade. As a modern industrial nation, the United States must participate actively in international trade. As basic trade theory emphasizes, the nation must import where its comparative advantage is low and export where it is high.

All evidence suggests that there are several important forest products in which the nation's comparative advantage is high. The decision to expand in these areas must be made by the forest products corporations and sources of investment capital on the basis of studies which are more product-specific and site-specific than those we are discussing here. The possible effectiveness of public policy in stimulating and enabling such an expansion is not entirely clear. Surely it is clear, however, that the objective of public policy should be to encourage such an expansion.

REFERENCE

U.S. Forest Service, 1979. "An Assessment of the Forest and Range Land Situation in the United States," review draft (Washington, D.C., U.S. Forest Service).

Discussion by John Ward

First, let me compliment David Darr on his presentation, "U.S. Imports and Exports of Some Major Forest Products: The Next Fifty Years." The paper is thorough and comprehensive and presents many insights into the U.S. position in world forest products trade. However, we at the National Forest Products Association (NFPA) would disagree with many of its conclusions regarding the solid wood sector of the industry.

Before commenting further, let me explain my role as director of the International Trade Committee for NFPA. Its function, and my responsibility, is to improve the climate, both within government and industry, for increased wood products exports. The committee was established because many key industry leaders believed the industry's potential for increased exports to be drastically underestimated. And the Forest Service is a major contributor to that underestimate.

At its 1979 annual meeting, NFPA's Board of Directors approved a major resolution evluating the 1980 Resources Planning Act (RPA) Draft Assessment and Program prepared by the Forest Service. The resolution highlighted the shortcomings of the latest RPA draft program, particularly its underutilization of public lands for forest productivity. Summing up, NFPA recommended to the secretary of agriculture that the nation's major timber goals should be to produce adequate supplies not only to reduce domestic inflation, but also to "enable the United States to become a net exporter of wood products."

Now why should NFPA, the national industry voice for the wood products portion of the forest products industry, take a major position that the industry can become a net exporter when it has been a net importer of wood products, essentially for the last twenty years? And why should it believe major increases in exports of wood products are possible when the Forest Service, the primary government agency for the industry as represented by Darr's paper, is forecasting minimal increases in such exports, and a continued huge net import position by the industry for the next twenty years?

The answer is complex, yet clear.

First, the Forest Service has built-in negative assumptions for supply in its economic models of the forest industry of the future. It inadequately estimates the impact of present and probable future forest management levels by the industry; thus, its forecasts understate supply and the amount of wood potentially available for export. Marion Clawon, of Resources for the Future, demonstrated the tendency of the Forest Service to underestimate future timber supply (Clawson, 1979). He graphed Forest Service forecasts since 1908, comparing forecast supply to what timber supply actually developed. In _every_ case, the Forest Service sharply underestimated the timber to be available in the future compared to what was subsequently grown. The current RPA forecast is reflected in Darr's predictions which assume that, over the next twenty years, the United States will not have enough timber; thus exports cannot be increased by much, and imports will be maintained at a high level.

Second, Darr's export forecast does not adequately consider growth in either total overseas markets' demand or in their demand for imports of ood. Over the next twenty years, world demand should grow faster than .S. domestic demand. And this demand will attract U.S. producers. Let e illustrate.

Domestic demand for softwood plywood will increase by 33 percent be-ween 1980 and 1990. However world demand will grow by 56 percent and orld import demand by 83 percent during this period.

Domestic demand for softwood lumber between 1980 and 1990 may grow aster than both world and world import demand; however, between 1990 and 000, world demand growth could be five times, and world import demand ten imes, greater than domestic growth.

Domestic demand for hardwood lumber is expected to grow very rapidly ver the next two decades. But world import demand growth could be 20 to percent greater.

Table 1 clearly illustrates this more dynamic potential in world mar-ets for selected products and compares past performance and future fore-asts (1970-2000) for growth in domestic, world, and world import demand.

Third, Darr's paper ignores the profit potential which the overseas arket offers and which will continue to draw more product into export. the short term many wood products are being sold at a premium in export arkets. Softwood logs exported from the West have been priced, for equiv-ent grade, at approximately 25 percent more than domestic logs over the ast five years. Softwood clear lumber exported from both the West and e South is highly valued product. Hardwood lumber sold overseas obtains 25 to 35 percent premium and hardwood veneer, 10 to 35 percent better ices compared to their domestic equivalents.

Table 1.   Percentage Increase of U.S. Domestic Versus World Versus World
            Import Demand Market Growth Compared for Selected Wood Products

Product and period	U.S. demand[a] (%)	World demand[b] (%)	World import demand[c] (%)
Softwood lumber			
1970-80	10	6	24
1980-90	35	14	20
1990-2000	2	9	17
Softwood plywood			
1970-80	25	18	100
1980-90	33	56	83
1990-2000	3	28	50
Hardwood lumber			
1970-80	0	12	88
1980-90	39	28	47
1990-2000	14	17	20

Note: "World demand" and "World import demand" excludes U.S. demands (domestic and import).

[a] Based on Forest Service projections.

[b] Based on FAO data, FAO and Stanford Research Institute forecasts and NFPA estimates.

[c] NFPA estimates

In the long term, producers will gain from the increased exports of more standard products such as dimension softwood lumber. Here improved profits will be obtained from increased volume, a broadened market, and the stability which exports can potentially contribute to the huge and damaging swings of the cyclical U.S. market.

Finally, Darr's or the Forest Service's projection of imports and exports, disregards the fact that the United States has a good competitive position in world markets today, and that its advantages in world markets are increasing for the future. America has a significant timber inventory in both softwood and hardwood. This inventory is more accessible, contains more valuable commercial species and is increasing at a faster rate than its major world competitors. The U.S. forest industry's demonstrated productive capacity in solid wood products is the largest in the world. And the infrastructure which supports it and the technological base it is built upon are unsurpassed.

To illustrate all of the foregoing, U.S. industry has maintained a 16 percent share of the world wood import market for the last ten years while the U.S. share of all merchandise exports had dropped from 23 percent to 17 percent during the same period. Certainly the U.S. wood products industry has demonstrated that it has a strong comparative advantage in world markets today.

Furthermore, America's competitors in the future should become less competitive. Canada's costs are increasing. Russia can only expand into Siberia. And Scandinavia cannot increase exports to meet growing world demand.

All in all, NFPA believes the United States is in a superb position for sharply increased exports in the world to come. It believes actual U.S. exports in the next twenty years will far exceed those forecast in Darr's paper and that actual U.S. imports, particularly by the year 2000, will be considerably below these estimated.

98

REFERENCE

Clawson, Marion. 1979. "Forests in the Long Sweep of History," Science
vol. 204 (June 15, 1979); RFF reprint 164.

PART II

THE EFFECT OF RESTRICTIONS UPON U.S. FOREST PRODUCTS TRADE

U.S.-CANADIAN LUMBER TRADE: THE EFFECT OF RESTRICTIONS

Darius Adams and Richard Haynes

The United States is the principal consumer of Canadian softwood lumber. In 1977 (the latest year for which official data are available) Canada produced 17.2 billion board-feet of softwood lumber, of which some 10.4 billion board-feet, or about 60 percent, was exported to the United States. Over the past decade, Canada has made steady inroads in U.S. markets. Canada's share of total U.S. new supply (domestic production plus imports) rose from 16.5 percent in 1968 to 27.7 percent in 1978. Much of this expansion has taken place in the traditional markets of the north central and northeastern regions. But southern markets have also accounted for a substantial portion of the growth in imports.

At present, U.S.-Canadian lumber trade is unencumbered by tariffs, quotas, or other trade restrictions. During the recent period of rapid import expansion, domestic producers have often suggested the need for protectionist measures, but no action has been taken. Raising trade barriers, when imports from Canada account for so large a fraction of U.S. consumption, would likely have significant effects both in the United States and Canada. This paper examines the market and resource impacts resulting from the hypothetical imposition of two such barriers: a 15 percent ad valorem import tariff on Canadian lumber and a fixed annual lumber import quota of 11 billion board-feet. The analysis of these measures is preceded by a brief review of the form and extent of U.S.-Canadian lumber trade.

Recent History of U.S.-Canadian Lumber Trade

The supply of Canadian lumber potentially available to U.S. consumers may be viewed as the residual of total Canadian lumber supply less Canadian consumption and exports to countries other than the United States. Thus, to understand the attributes of Canadian-U.S. lumber trade it is instructive to examine first the characteristics of these latter three flows

Canadian Lumber Production

Total Canadian lumber production and its disposition among domestic consumption, exports to the United States, and exports to other countries are illustrated in figure 1. During the decade from 1968 to 1978, total production increased some 68 percent. The figure clearly indicates the expanding importance of U.S. markets, with exports to the United States increasing by nearly 105 percent between 1968 and 1978. Domestic consumption increased about 29 percent during the same decade while exports to countries other than the United States remained roughly stable.

Figure 2 shows Canadian softwood lumber production broken down by major producing regions. While output has expanded in virtually all provinces, the most rapid increases have taken place in British Columbia particularly in the interior region. Sparked by extensions of the Canadian rail system and development of small log processing techniques, interior British Columbia has become the largest producing region in Canada, replac-

Figure 1.   Canadian softwood lumber production and its disposition,
            1950–1978.

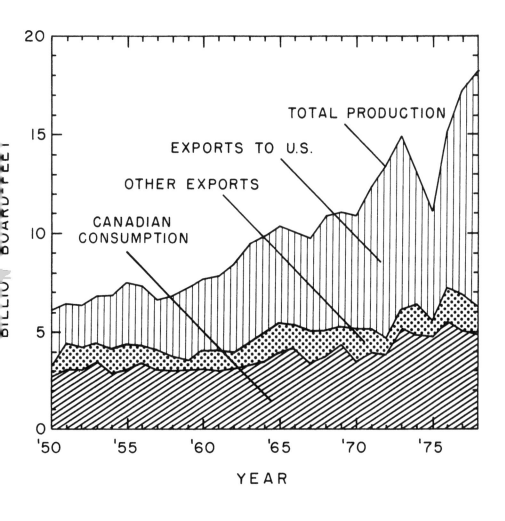

Figure 2. Canadian softwood lumber production by region, selected years, 1950-1978.

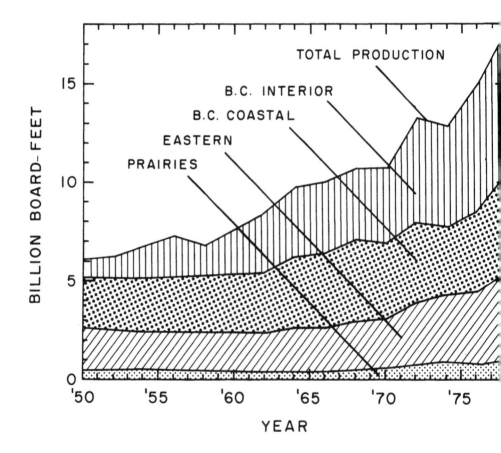

ing the coastal region as the principal supplier of exports to the United
States. Interior and coastal British Columbia differ markedly in terms of
production technology employed, markets, and species composition of output.
Interior production is based largely on smaller logs and modern high-volume
production techniques. Coastal mills utilize larger old-growth timber and
more traditional sawing methods. In 1978 interior mills shipped 21 percent
of their production to Canadian destinations, 78 percent to the United
States, and 1 percent to the United Kingdom and Pacific markets. Coastal
mills, on the other hand, shipped 23 percent to Canadian markets, 48 percent to the United States, and 29 percent to the United Kingdom and Pacific
points. Finally, lumber production in the coastal region is concentrated
in hemlock (59 percent), red cedar (17 percent) and Douglas fir (16 percent). Interior mills produce primarily spruce (66 percent), pine and true
firs (16 percent), and Douglas fir (10 percent).

The rapid expansion in Canadian production and exports to the United
States since the mid-1960s has been possible largely because Canadian
lumber production costs have not risen as rapidly as those in the major
U.S. producing regions. As a result, Canada has maintained and improved
its competitive advantage in U.S. lumber markets. Figure 3 compares softwood lumber production costs (wood plus nonwood inputs) for Canada and the
United States, Douglas fir and southern regions. During the 1950s and
early 1960s, lumber production in the South declined sharply (1962 southern
production was 41 percent below the 1950 level) as the wide cost gap in
Figure 3 suggests. Both Canada and western U.S. regions captured portions
of the markets lost by the South. Beginning in the late 1960s, strong
periodic surges in U.S. housing activity drove up the demand for lumber.

Figure 3. Lumber production costs in Canada and the Douglas fir and southern U.S. regions, 1950-1976.

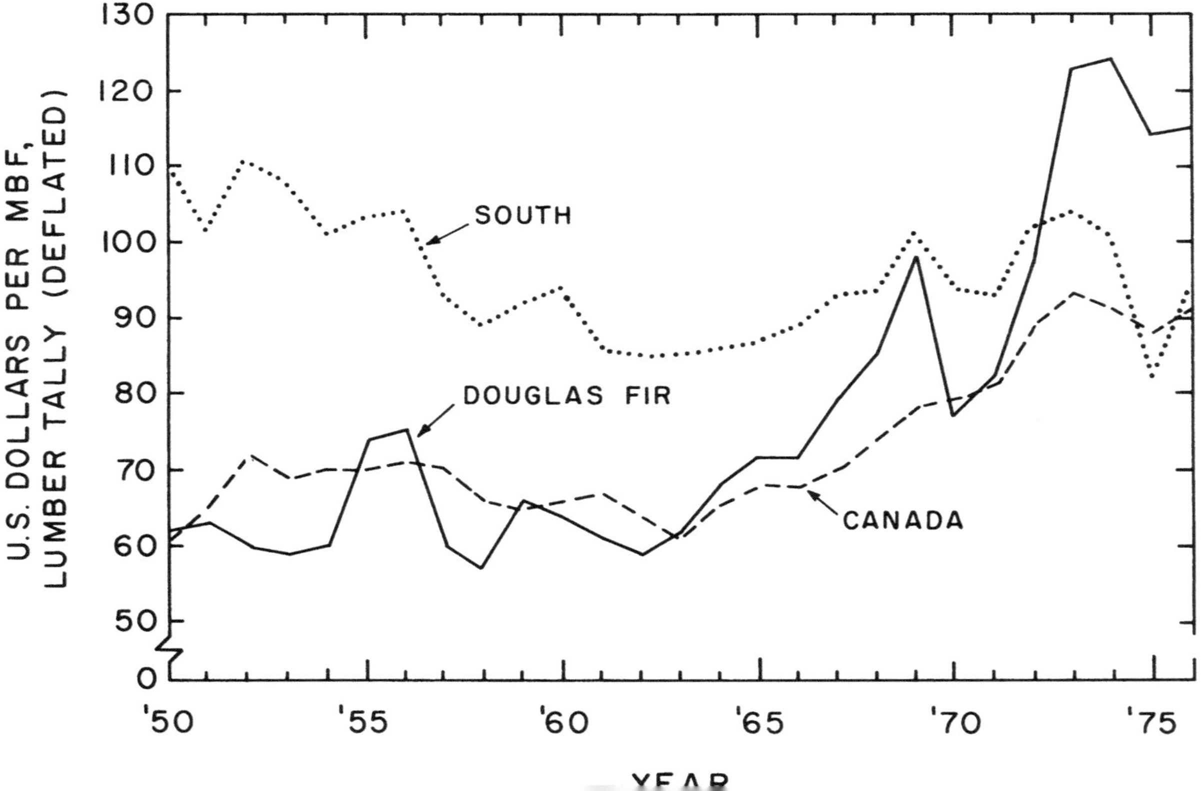

t about the same time, U.S. Pacific Coast lumber producers began to ex-
erience increasing difficulties in acquiring stumpage, owing to sharply
eclining private inventories and stabilization of public harvests. Stum-
age prices began to rise rapidly, pushing lumber production costs to, and
ventually above, the level of costs in the South. Canadian costs also
egan to rise in the mid-1960s, but less rapidly than those in the Douglas
ir region. The principal source of these increases has been growing tim-
er harvesting and log transport costs. Southern costs showed no signifi-
nt trend. As a result of these relative cost movements, Canada, and to
lesser extent the U.S. South, captured the bulk of U.S. demand increases
ince the late 1960s. Production in western U.S. regions either remained
able or declined slightly during this period.

A final attribute of Canadian lumber production of considerable im-
rtance in determining export behavior is the responsiveness (or elas-
city) of supply. In general the elasticity of export supply is directly
lated to the elasticity of aggregate supply. For importing countries
ich as the United States, where imports are a large part of total consump-
ion, the more elastic is the supply of imports (which is essentially iden-
ical to Canadian export supply) the less marked will be any price movement
sulting from shifts in demand, ceteris paribus. Stated differently, a
igh import supply elasticity tends to provide a greater buffer against
ice movements when domestic demand shifts.

Supply elasticity in the long-run (over a period of several years) de-
ends on how rapidly unit production costs change at a given level of capa-
ity and how rapidly capacity (and hence investment) adjusts to changing
nditions of profitability. Assuming a competitive Canadian softwood lum-

ber industry operating so as to equate price and marginal unit production cost, some indication of the elasticity of Canadian supply can be obtained from an examination of the historical relation between price and total production. These data are plotted in figure 4. There appear to be four major groupings of price-quantity points. During the period from 1950 through 1963, production generally expanded despite stable or declining prices. This was a transitional period, not unlike the early years of the softwood plywood industry in the United States in which producers were adjusting to new technology and new opportunities for production (interior British Columbia production quadrupled during this period exceeding coastal production for the first time). The period from 1964 through 1969 saw the first major price rise of the post-World War II period. Production also increased but to a limited extent. A linear function fit to these data points yields an average price elasticity of about 0.40. Assuming stable supply and shifting demand, this elasticity might be interpreted as the elasticity of supply. A third period, from 1970 through 1974, corresponds to a major cycle in U.S. housing starts. Supply appears to have been somewhat more responsive to price than during the 1964-69 period. A linear estimate yields an elasticity of about 1.0. The final group of points, corresponding to the most recent U.S. housing surge, appears still more elastic with a linear estimate of 1.6

While this simple analysis does suggest that Canadian lumber supply is fairly price responsive and that its elasticity may have increased over the past two decades, the specific elasticity estimates may, of course, be misleading. A recent study by Adams and Haynes (in press), however, provides consistent econometric estimates of supply elasticties for Canada

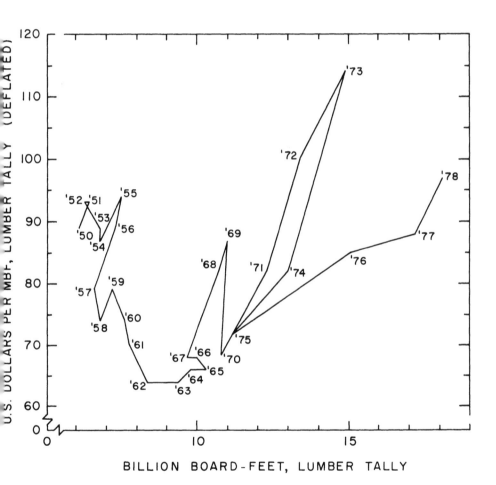

Figure 4. Relationship of Canadian lumber production and price, 1950-1978.

and the major western and southern U.S. producing regions. These elasticities (computed at sample period means) are tablulated below:

Region	Adams and Haynes softwood lumber supply elasticity
Douglas fir	.21
Ponderosa pine	.60
California	.23
Rockies	.35
Southcentral	.79
Southeast	.31
Canada	.47

From these results, total Canadian supply is seen to be more elastic than supply in all but the two most rapidly growing U.S. regions. It is the elasticity of export supply, however, that is of concern in U.S. markets. As previously noted, this may be defined as export supply = total supply − domestic consumption − non−U.S.−exports. After appropriate adjustments for these latter two elements, Adams and Haynes found the elasticity of Canadian export supply to be approximately 0.89 (computed at sample period means). It is commonly suggested that Canadian imports are the first supply element to increase during upswings in U.S. lumber markets and the first to fall during declining markets. This volatile behavior, and the price buffering effects implicit in it, is consistent with the above findings which suggest that Canadian imports are the most elastic component of total U.S. lumber supply.

Canadian Consumption and Off-Shore Exports

As indicated in figure 1, Canadian consumption of softwood lumber has grown steadily over the past three decades, rising from roughly 3 billion board-feet in the early 1950s to about 5 billion board-feet in the late 1970s. In recent years domestic uses have consumed about one-third of total Canadian production. About half of domestically consumed lumber originates in British Columbia (slightly more from the interior than the coast) with the remainder coming from other provinces (predominantly Ontario and Quebec). Imports, at 0.5 percent of total production, are negligible. Underlying the historical growth in consumption has been a continued expansion in Canadian housing activity. Between the early 1950s and the late 1970s housing starts rose from 92,000 to nearly 230,000 annually. Over the same period the total real value of Canadian construction put-in-place increased nearly threefold. The Canadian experience is thus quite unlike the recent history of U.S. lumber consumption, in which there has been no consistent trend either in housing activity or lumber use over the past thirty years.

Canadian lumber demand, like that in the United States, appears to be relatively unresponsive to changes in price in the short term (one to two years). Econometric efforts to estimate Canadian demand elasticity by Manning (1975) and the authors support this conclusion. The important effect of a low domestic demand elasticity in the current context is to reduce the elasticity of export supply. Intuitively, this arises because a less elastic domestic demand does not decline as rapidly as when prices rise, thereby freeing less for exports, not does it expand as rapidly when prices fall, leading to a smaller reduction in export supply volumes.

Canadian exports to countries other than the United States have shown considerable fluctuation but no significant trend over the past three decades, averaging about 1.5 million board-feet per year. The principal destinations are Japan, the United Kingdom, Australia, and other Commonwealth countries. About 90 percent of these shipments originate in coastal British Columbia. Non-U.S. exports exhibit little price responsiveness, suggesting inelastic import demand in receiving countries (econometric evidence from Manning (1975), and work by the authors support this finding). Low elasticity in this case may result from trade restrictions, such as Japan's effective import quota on Canadian lumber.

Exports to the United States

In recent years about 79 percent of Canadian exports to the United States originated in British Columbia, 4 percent in the prairie provinces, and 17 percent in the eastern provinces (largely Ontario and Quebec). Within British Columbia, about 72 percent of exports come from the interior and the remainder from the coast. Reflecting this geographic distribution of points of origin, U.S. imports from Canada have consisted of some 54 percent spruce, 16 percent pine, 10 percent hemlock, 8 percent Douglas fir, and 12 percent of other species (predominantly red cedar).

Within the United States, imports from Canada are a large and growing component of total consumption. Table 1 indicates historical trends in the sources of softwood lumber consumed in the United States. Between 1950 and 1978, imports from Canada quadrupled, rising from 9 percent to nearly 29 percent of total consumption. As the volume of imports has risen the destination of import shipments within the United States has also changed. Estimates of the distribution of Canadian shipments to U.S.

able 1. Total Apparent U.S. Consumption of Softwood Lumber and Sources
         of Supply, for Selected Years

(illion board-feet)

| ear | Supply Source | | | Total apparent consumption |
	U.S.[a]	Canada	Other	
950	30.2	2.9	.2 [b]	33.3
955	29.2	3.2	.1	32.5
960	26.0	3.6	-	29.6
965	28.3	4.9	-	33.2
970	26.0	5.7	-	31.7
975	25.0	5.7	-	30.7
978	29.2	11.8	.1	41.0

[a]U.S. production less exports.

[b]Dashes denote zero value.

egions are shown in table 2. These data indicate that the traditional
ortheastern and north central markets have gradually declined in impor-
ance while the rapidly growing southern states have increased drama--
ically. The northeastern and southern regions are primarily waterborne
arkets served by coastal British Columbia. The north central region is
rimarily a rail market served by interior British Columbia and the east-
rn provinces.

Analysis of Restrictions

Canadian-U.S. lumber trade in the period after World War II has been
ree of restrictions, with the exception of a 25-cent tariff per 1,000

Table 2. Distribution of Lumber Imports from Canada by U.S. Regions,
          Expressed as a Percentage of Total Imports

	Region of Destination (%)[a]					
Year	Northeast	Northcentral	South	Rockies	Southwest	Northwest
1960[b]	41	52	– [c]	–	2	5
1965[b]	35	47	11	–	2	5
1970	32	40	21	–	2	5
1975	26	43	24	1	1	5
1978	20	40	26	3	4	7

Source: Statistics Canada. Special tabulations of Canadian exports
by U.S. Census region of destination; and author's estimates.

[a]Regional definitions:

Northeast: all states north and east of Pennsylvania
North central: Ohio, Indiana, Illinois, Missouri, Iowa, Nebraska,
                North and South Dakota, Kansas, Minnesota, Wisconsin,
                and Michigan
South: all other states east of the Rockies and not included in
        northeast or north central
Rockies: Montana, Idaho, Utah, Wyoming, and Colorado
Southwest: California, Nebada, Arizona, and New Mexico
Northwest: Oregon and Washington

[b]Author's estimates.

[c]Dashes denote zero value.

board-feet imposed following World War II and finally eliminated during

the Kennedy round of trade negotiations in the early 1960s. The tariff

was so small, however, as to pose no significant trade barrier. Despite

the absence of restrictions in the recent past, an understanding of the

potential effects of trade barriers provides insights into the functionin of the North American softwood lumber economy and may prove useful in the evaluation of suggestions for the imposition of such barriers that arise from time to time.[1] In this section we examine two of the most common types of trade restrictions, an ad valorem import tariff and an import quota. Each restriction is imposed in turn and the resulting impacts on the United States and Canada projected over the period from 1980 to 2000. Projections were made by means of a spatial equilibrium model of U.S. forest products markets developed by Adams and Haynes (in press). Impacts on production, consumption, prices and trade as well as effects on the U.S. forest resource base are presented.

Assumptions and Conditions of the Analysis

The projections of future market activity made with the Adams-Haynes model are conditional on assumptions regarding the future levels of several economic variables not explained by the model.

Canadian consumption and exports to countries other than the United States. Canadian softwood lumber consumption was assumed to rise at approximately 1.2 percent per year over the next two decades (about half of the 1950-78 growth rate). Off-shore exports rise by 1.0 billion feet. Both assumptions were suggested by recent Canadian studies.

Production costs in the United States and Canada. Based on assumptions regarding trends in labor productivity in logging and manufacturing, wage rates, machinery costs and fuel prices, nonwood lumber production

---

[1] As noted in the introduction, expansion of U.S. lumber imports in the 1960s generated some interest among industrial groups in the imposition of import duties. More recently, a member of the Oregon congressional delegation proposed the acceleration of reforestation measures on the national forests funded by duties on imported forest products.

costs were projected for all U.S. producing regions. For the Douglas fir region, for example, projections of deflated nonwood costs increase by 2.25 percent per year between 1980 and 2000, while they grow at 2.11 percent in the South. Wood costs (stumpage prices) are determined in the model and vary between simulations. Deflated Canadian total costs (including wood) were projected to rise at 2.19 percent.

Transportation costs. All transportation costs within the United States and between Canadian and U.S. points were assumed to remain constant in real terms at 1977 levels.

Exchange rates. Despite the recent decline in the U.S.-Canadian exchange rate (Canadian dollars becoming cheaper in terms of U.S. dollars) it was assumed that the rate would return to its long term historical level of roughly 1.0 and remain at that level for the next two decades.

Ad valorem Tariff

The market effect of an ad valorem tariff is to drive a wedge between prices in the importing and exporting countries. Prices, total supply, and exports from the exporting country decline, while prices and production in the importing country rise and consumption declines. Figure 5 illustrates these results. Canadian demand is shown as perfectly inelastic. Without the ad valorem tariff, exports to the United States are the volume $S_{wo} - D_c$. Imposition of a tariff shifts the excess Canadian supply function upward throughout its length and increases its slope raising prices in the United States for equivalent import volumes. The new trade equilibrium occurs as a result of the lower volume, $S_w - D_c$.

Using the Adams-Haynes model, a 15 percent ad valorem tariff on Canadian lumber imports was simulated over the 1980-2000 period. Impacts are

Figure 5.   Effects of <u>ad valorem</u> import tariff on Canadian softwood lumber
            upon U.S.-Canadian lumber trade.

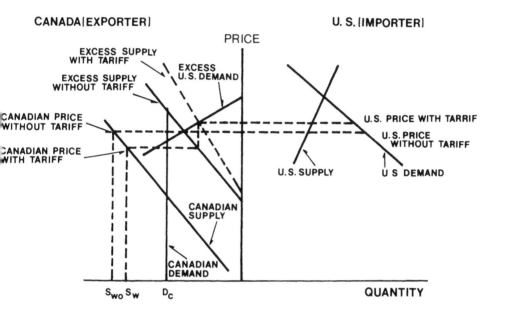

measured as the difference between projections with the tariff imposed and

a "base" projection without the tariff.   All other assumptions and con-

ditions in the two projections were the same.

   Aggregate projection results for the lumber market are shown in table

3.   Under the tariff, imports from Canada decline by 4,439 million board-

feet by 1990 and 5,598 million board-feet by 2000.   U.S. production rises

as prices increase but by less than the decline in imports.   Thus, total

U.S. softwood lumber consumption falls by 1,507 million board-feet in

1990 and by 2,131 million board-feet in 2000.   Because other demands for

Table 3. Production, Consumption, and Price Impacts of 15 Percent Ad Valorem Import Tariff on Canadian Softwood Lumber

Softwood lumber (MBF)	1985 Without tariff	1985 With tariff	1990 Without tariff	1990 With tariff	1995 Without tariff	1995 With tariff	2000 Without tariff	2000 With tariff
U.S. production	32,029	33,657	31,416	34,133	30,133	33,178	29,424	32,890
Canadian imports	9,958	6,822	12,665	8,226	13,487	8,460	13,744	8,146
Other imports	75	75	75	75	75	75	75	75
Exports	1,485	1,485	1,600	1,600	1,550	1,550	1,500	1,500
U.S. consumption	40,577	39,070	42,556	40,834	42,146	40,163	41,743	39,612
Canadian production	16,941	13,806	20,065	15,626	21,537	16,510	22,444	16,846
U.S. softwood lumber producer price index (67 = 100)	175	188	200	214	212	232	225	246
Canadian lumber price (U.S.$/MBF)	135	130	155	147	165	159	179	172
Douglas fir region stumpage ($/MBF)	88	94	104	111	95	105	87	103
Southern region stumpage ($/MBF)	72	76	85	94	94	105	104	118
U.S. softwood plywood consumption (million square feet)	20,738	20,554	21,799	21,526	22,001	21,654	21,998	21,566

anadian lumber were assumed to be perfectly inelastic, Canadian produc-
ion falls by the amount of the reduction in exports to the United States.
oftwood lumber prices in the United States, as a result of the supply con-
triction, rise by 7 percent in 1990 and by 9 percent in 2000. Canadian
rices fall, by 5 percent in 1990 and 4 percent in 2000.

Because of interaction in stumpage markets tariff impacts are not con-
ined to the lumber industry. Increases in stumpage prices influence the
ost of producing plywood, shifting the plywood supply function upward in
rice-quantity space. The net result, as indicated in table 3, is a modest
eduction in U.S. plywood consumption, reaching -2 percent by 2000. Simi-
arly, in the pulp and paper industry, increased stumpage demand for lumber
roduction and increased lumber output shift the relative costs and availa-
ility of roundwood pulpwood and residues. Table 4 shows the resulting
anges in the national mix of roundwood and residues used in pulping. By
00 residue use has increased from 29 to 33 percent of wood fiber input
der the tariff, freeing some 155 billion cubic feet of softwood roundwood
nually for other uses.

At the regional level, softwood lumber production under the tariff
creases throughout the United States (see table 5). Because of the
eater elasticity of lumber supply and superior competitive position of
e South, production increments and expansion in market share are great-
t to that region. By 2000, some 59 percent of additional U.S. lumber
roduction under the tariff originates in the south, 20 percent in the
uglas fir region, 10 percent in the Rockies, and the remainder in Cali-
rnia and the ponderosa pine regions. The tariff reduces but does not
iminate the decline in Douglas fir and southern region production, par-

Table 4. Percentage Distribution of Softwood Roundwood and Residue Consumption in Pulping With and Without a 15 Percent Tariff on Canadian Lumber Imports to the United States

| | 1990 | | | 2000 | |
	Without tariff	With tariff		Without tariff	With tariff
Roundwood	66	64		71	68
Residue	34	36		29	32

Table 5. Softwood Lumber Production by U.S. Region and Imports and Percentage of U.S. New Supply, With and Without 15 Percent Ad Valorem Tariff

(million board-feet)

| | 1990 | | | | 2000 | | | |
Region	Without tariff	(%)	With tariff	(%)	Without tariff	(%)	With tariff	(%)
Douglas fir	5,545	(13)	5,797	(14)	4,112	(10)	4,798	(12)
Ponderosa pine	2,916	(6)	3,061	(7)	3,095	(7)	3,240	(8)
California	5,570	(13)	5,773	(13)	5,491	(13)	5,740	(14)
Rockies	5,437	(12)	5,837	(14)	5,683	(13)	6,030	(15)
South	10,665	(24)	12,383	(29)	9,658	(22)	11,695	(28)
North	1,282	(3)	1,282	(3)	1,387	(3)	1,387	(3)
Total	31,416	(71)	34,133	(80)	29,426	(68)	32,890	(80)
imports	12,740	(29)	8,301	(20)	13,819	(32)	8,221	(20)
New Supply	44,156	(100)	44,434	(100)	43,245	(100)	41,111	(100)

Note: New supply = domestic production + imports. Percentages are given in parentheses.

:icularly offsetting rising regional stumpage costs with an increase in
'effective" lumber demand.

Increased lumber production in U.S. regions augments the derived de-
and for stumpage and leads to higher stumpage prices. As shown in table
$, Douglas fir regions are 18 percent higher by 2000 while southern region
rices are 13 percent higher. In the absence of the tariff, Douglas fir
egion prices decline after 1990 as a result of stabilizing U.S. lumber
emand and the rapid exodus of lumber capacity.[2] The tariff attenuates
his decline, by retaining more capacity in the region, but does not elimi-
ate it.

Effects of the tariff on total timber harvest by region are shown in
able 6, and inventory levels on private lands for selected regions are
hown in table 7. Harvest increases in all regions but by substantially
ess than the roundwood equivalent of the increases in lumber production.
his is caused by the partially offsetting decline in harvests for plywood
nd pulp production. By 2000, total U.S. harvest is only 360 million cubic
eet higher under the tariff, an increase of less than 3 percent. Because
f the insensitivity of public harvests to price fluctuations, virtually
ll of this increase comes from private lands. Table 7 indicates that im-
acts on the rapidly declining private inventory in the Douglas fir region
re relatively minor, since the plywood and pulpwood offset is substantial.
n the south, however, expanded lumber output under the tariff converts a
odest inventory increase to a nearly equivalent decline.

---

[2]Projections of housing activity decline between 1990 and 2000 owing
o demographic trends and with them, lumber demand.

Table 6. U.S. Softwood Timber Harvest by Region With and Without a 15
          Percent Ad Valorem Canadian Lumber Import Tariff

(million cubic feet)

Region	1990 Without tariff	1990 With tariff	2000 Without tariff	2000 With tariff
Douglas fir	2,477	2,496	2,237	2,277
California	905	928	875	901
Rockies and ponderosa pine	1,625	1,696	1,711	1,778
South	5,689	5,884	6,239	6,466
North	891	891	1,011	1,011
Total	11,587	11,895	12,073	12,433

Table 7. Private Softwood Growing Stock Inventories for Selected U.S.
          Regions With and Without a 15 Percent Ad Valorem Canadian
          Lumber Import Tariff

(million cubic feet)

Region	1990 Without tariff	1990 With tariff	2000 Without tariff	2000 With tariff
Douglas fir	24,630	24,324	21,399	20,684
California	14,257	13,853	12,758	12,019
South	96,180	94,966	97,697	93,811

Presumably, a tariff of the sort investigated here would be adopted to benefit U.S. industry, or consumers, or both. The theoretical analysis in figure 5 suggests that consumers would probably not benefit since consumption declines while prices rise. The extent to which industry benefits depends on the relative increase of gross revenues and production costs (primarily stumpage prices). Using the demand and supply relations in the Adams-Haynes model, estimates of consumer and producer surpluses in the lumber market alone were obtained with and without the tariff. As indicated in the following tabulation, increments in producer surplus exceed losses in consumer surplus by 1990 with an even wider difference by 2000. Of course, a complete cost-benefit accounting would require consideration of impacts on stumpage producers and the producers and consumers of substitute and complementary goods.

(billion U.S. $)

	Lumber market consumer surplus			Lumber market producer surplus		
	Without tariff	With tariff	Change	Without tariff	With tariff	Change
1990	7.024	6.460	-.564	3.923	4.627	+.704
2000	7.089	6.373	-.716	3.894	4.866	+.972

Import Quota

Under an import quota, U.S. demand for imports would be restricted to volumes no larger than the arbitrary quota level, which would in turn be less than the volume traded under free market conditions. Like the import tariff, a quota increases price and domestic production while lowering consumption in the importing country. In the exporting country both price and production are reduced. Figure 6 illustrates these conditions. The

Figure 6. Effects of import quota on Canadian softwood lumber upon U.S.-
Canadian lumber trade.

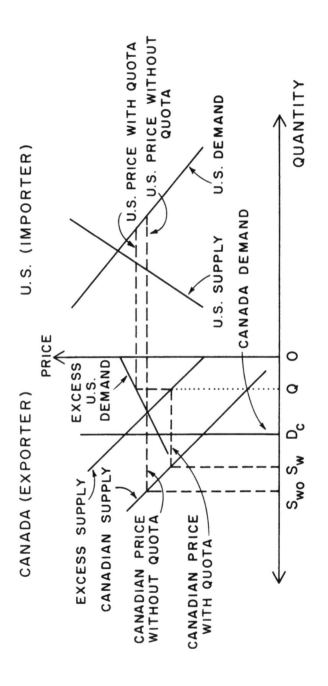

uota level, Q, is imposed on the excess Canadian supply of lumber. In

he absence of the quota, $S_{wo}-D_c$ would be exported to the United States.

'ith the quota, the export volume falls to $Q = S_w-D_c$.

For the purposes of this analysis, a quota level of 11.0 billion

oard-feet was set, roughly equivalent to the average 1977-78 import level

rom Canada. All other assumptions and conditions in the simulation are

s previously outlined. Aggregate results are shown in table 8. Because

he simulation begins at trend levels (rather than actual) in 1979, the

uota does not become effective until about 1988 in the simulation. By

990, U.S. softwood lumber production has risen only 2 percent and con-

umption declined by about 2 percent. Since the quota allows a larger im-

ort volume than did the tariff, impacts by 2000 are not as great as those

f the tariff. U.S. production by 2000 is up 6 percent and consumption

own by only 2 percent. Lumber prices in the United States rise and those

n Canada fall as expected from the analysis. Plywood production and the

istribution of wood input to pulping between roundwood and residues are

nly modestly altered.

The most significant differences between the quota and tariff res-

rictions lie in their impacts on producing and consuming regions. Table

shows regional lumber production and shares of new supply (new supply =

omestic production + imports) under the quota (compare table 5). Note

hat under the quota nearly 87 percent of the total U.S. production in-

rement is concentrated in the Pacific Coast regions (Douglas fir and

alifornia). In contrast, 73 percent of the production increment due to

he tariff was realized by ponderosa pine (4 percent), the Rockies (10

ercent), and the South (59 percent). In the consuming regions, total

Table 8. Production, Consumption and Price Impacts of an 11-Billion-Board-Foot-Quota on U.S. Softwood Lumber Imports from Canada

126

	1985		1990		1995		2000	
	Without quota	With quota	Without quota	With quota	Without quota	With quota	Without quota	With quota
Softwood Lumber (MBF)								
U.S. production	32,029	32,029	31,416	32,189	30,133	31,686	29,424	31,180
Canadian imports	9,958	9,958	12,665	11,000	13,487	11,000	13,744	11,000
Other imports	75	75	75	75	75	75	75	75
Exports	1,485	1,485	1,600	1,600	1,550	1,550	1,500	1,500
U.S. consumption	40,577	40,577	42,556	41,664	42,146	41,211	41,743	40,755
Canadian production	16,941	16,941	20,065	18,400	21,537	19,050	22,444	19,700
U.S. Softwood lumber producer price index ('57=100.0)	175	175	200	205	212	221	225	237
Canadian lumber price (U.S.$/MBF)	135	135	155	145	165	161	179	177
Douglas fir region stumpage ($/MBF)	88	88	104	109	95	109	87	110
Southern region stumpage ($/MBF)	72	72	85	86	94	95	104	105
U.S. softwood plywood consumption (million square feet)	20,738	20,738	21,799	21,734	22,001	21,784	21,998	21,809

Table 9. Softwood Lumber Production by U.S. Region and Imports and Percentage of U.S. New Supply With Import Quota

(million board-feet)

Region	1990		2000	
Douglas fir	5,828	(13)	5,371	(13)
Ponderosa pine	2,946	(7)	3,105	(7)
California	5,816	(13)	5,945	(14)
Rockies	5,474	(13)	5,459	(13)
South	10,843	(25)	9,913	(24)
North	1,282	(3)	1,387	(3)
Total U.S.	32,189	(74)	31,180	(74)
Imports	11,075	(26)	11,075	(26)
New Supply	43,264	(100)	42,244	(100)

Note: New supply = domestic production + imports. Percentages are
given in parentheses.

consumption declines under both forms of restriction but the regional
distribution of impacts is markedly different as table 10 indicates.

Under the quota the bulk of the reductions occur in the West, while
under the tariff 67 percent of the decline is concentrated in the East.
Of course, the extent of the consumption decline is quite different in
the two cases (2.1 billion with the tariff and about 1.06 billion with
the quota).

The causes of these differential impacts lie in the changing patterns
of regional comparative advantage in lumber production. At the beginning
of the projection period (1980) the South, benefiting from lower wood and
nonwood costs, is in a better position to respond to increases in demand

Table 10. Percentage Distribution of Year 2000 Reduction in Lumber Consumption Caused by Trade Restriction

	Region					
Restriction	Northwest	Southwest	Rockies	Northcentral	Northeast	South
Quota	19	57	-	6	6	12
Tariff	7	20	5	26	11	30

Table 11. U.S. Softwood Timber Harvest by Region Under a Canadian Lumber Import Quota

(million cubic feet)

Region	1990	2000
Douglas fir	2,500	2,317
California	938	927
Rockies and ponderosa pine	1,634	1,705
South	5,713	6,252
North	891	1,011
Total	11,676	12,212

than regions in the West. As production increases during the 1980s, however, southern stumpage prices rise rapidly with or without trade restrictions. By 1990, the South's production advantage has been markedly reduced as a result, and production declines steadily to 2000. The Pacific Coast regions, on the other hand, face high wood and nonwood costs at the outset of the projection. Margins for profit are low and capacity declines rapidly in these regions during the 1980s. This loss reduces pressure on stumpage prices and gradually improves the competitive position of the regions during the 1990s. Thus import restrictions such as the tariff, imposed early in the projection period, yield larger initial increases in southern production. What additional production is obtained from the Douglas fir and California regions under the tariff restriction only serves to exacerbate their rates of stumpage cost increase. Restrictions such as the quota, which becomes effective later in the projection, elicit larger responses from the Pacific Coast regions. Production continues to decline in these regions but at a slower rate than in the absence of import restrictions.[3]

Differences in the distribution of consumption impacts between tariff and quota restrictions derive from regional patterns of production expansion. Imports from Canada flow primarily to the northeastern and north central regions, so that both restrictions initially create consumption "gaps" in these regions. In the tariff case, the Rockies expands shipments to the north central region by redirecting shipments that would otherwise have gone to the South. The South expands shipments to the northeast by redirecting intraregional trade. At the same time, some ponderosa pine region production is shifted from the Southwest to the South. Since lumber supply in the South, Rockies, and ponderosa pine regions is not as elastic as that from Canada (before the tariff), the net result is a major reduction in eastern consumption. California and the Douglas fir

---

[3]Simulation of an import quota of 8.75 billion board-feet leads to a distribution of consumption and production impacts virtually identical to that found for the 15 percent tariff. This lower quota becomes effective in the simulation in 1983 (in contrast to 1988 for the 11 billion-board-foot quota).

region are precluded from expanded participation in eastern markets for reasons of cost as discussed above. In the quota case, it is the South that faces competitive disadvantages. Thus, it is primarily shipments from the Rockies that move into the North, while ponderosa pine production is almost completely directed to the South from its former primary destination in the Southwest. The Pacific Coast regions are unable to fully substitute for this latter loss, yielding a large consumption impact in the Southwest. Eastern impacts are relatively minor.

Tables 11 and 12 show the harvest and private inventory impacts of the quota. Changes relative to the tariff results are as expected in light of the preceeding discussion of the differences in lumber market activity.

Finally, welfare impacts of the quota related solely to the lumber market are shown in the following tabulation. Unlike the tariff, increments in producer surplus do not consistently exceed the losses in consumer surplus. Once again this results from the differences in regional distribution of shifts in production and consumption between the quota and the tariff.

(billion 1967 $)

	Lumber market consumer surplus			Lumber market producer surplus		
	Without quota	With quota	Change	Without quota	With quota	Change
1990	7.024	6.732	-.292	3.923	4.141	+.218
2000	7.089	6.757	-.332	3.894	4.391	+.497

Table 12. Private Softwood Growing Stock Inventories for Selected U.S.
Regions Under a Canadian Lumber Import Quota

(million cubic feet)

Region	1990	2000
Douglas fir	24,601	20,769
California	14,221	12,192
South	96,153	97,380

Conclusion

In the absence of trade restrictions, imports of Canadian lumber are
projected to grow to nearly 33 percent of total U.S. consumption by the
year 2000. Analysis of a tariff and a quota designed to limit this potential growth in imports suggests that it is the timing of the imposition
of the restriction rather than its form that is central in determining its
impacts over the next two decades. In the case of the tariff, and in
general for either a tariff or quota imposed early in this decade, reductions in consumption would likely be concentrated in the eastern United
States while expanded domestic production would come primarily from the
South. Estimated welfare changes in the lumber market alone suggest that
producer gains would outweigh consumer losses. For the quota, and also
a tariff imposed late in this decade, consumption and production impacts
would be concentrated in the West. Lumber market welfare changes, given
different geographic distribution of consumption and production shifts
relative to the tariff, show initial net losses (by 1990) changing to
modest net gains by 2000.

Both forms of restrictions lead to higher rates of stumpage price growth in the United States, as increased lumber output augments the demand for timber. This would likely provide additional benefits to producers of stumpage but at the same time represents added costs for producers of plywood and wood pulp. Analysis indicates a modest decline in U.S. plywood production and consumption and a shift in the mix of wood fiber input to pulping from roundwood toward a higher proportion of residues as a result. Higher stumpage prices also lead to higher levels of timber harvest. On the assumption that public timber harvest is only modestly responsive to price, the bulk of the cut increase must come from private lands. In the West this leads to further depletion of an already declining private inventory. In the South, it slows down, or in the case of the tariff, reverses, the rate of private inventory accumulation.

REFERENCES

Adams, Darius M., and Richard W. Haynes. (In press). "The 1980 Softwood Timber Assessment Market Model: Structure, Estimation, and Policy Simulations," Forest Science Monograph.

Manning, Glenn H. 1975. "The Canadian Softwood Lumber Industry: A Model," Canadian Journal of Forest Research, vol. 5, no. 3, pp. 345-351.

Discussion by A. Clark Wiseman

Authors Adams and Haynes are to be lauded for their intrepid venture into the highly uncertain realm of long-range forecasting. It is an exercise which is more difficult to carry out than to critique, notwithstanding the authors' having obtained five-significant-figure accuracy twenty years hence from some rather "iffy" two-significant-figure elasticities and growth rates. I will make three general comments, to be interpreted--I hope--as constructively critical.

First, one would like to know more about the policy relevance of the two hypothetical restrictions investigated in the paper. Why a 15 percent tariff, rather than more or less? More important, the size of the hypothetical quota seems to have little to recommend it as a policy worthy of study. We are told--well into the paper--that the $11 billion board-feet was the average 1977-78 import level from Canada. As such, it could hardly be expected to have the initial restrictive effect of the tariff: this is indeed "discovered" by the analysis. All tariffs have a theoretical equivalent quota which will have the same initial import restricting effect. It would seem to be more interesting and relevant to compare a given tariff with its quota equivalent to see which is most damaging over time.

Use of the word "damaging" indicates my refusal to accept the finding
of the paper that there is a tendency for there to be a net welfare gain
to the United States from either restriction. This leads to the second
general point. It is a well-established finding of international trade
theory that, in competitive static analysis and ignoring government revenue
gains, a tariff or quota results in a net welfare loss to the country oppo-
sing the restriction unless certain conditions prevail. Either there must
be externalities that result in resource misallocation under free trade or
else there must be a "terms of trade" effect so great that import price is
driven down to a point where the net price (including tariff) to buyers in
the tariff-imposing country is actually less than without the restriction.
The externalities argument is not mentioned in the paper, and the tables
clearly show the U.S. lumber price being higher under the tariff or quota.

It is inconceivable, given the conditions postulated, that producers
can gain more than buyers loose. By the year 2000, U.S. production is
projected in the paper to be about two-thirds the level of consumption.
Therefore, as a first approximation the price-increasing effect of the
restriction will yield two-thirds as much gain to producers as loss to
consumers. The paper's findings may result from the use of a technical
measurement that differs from the accepted Marshallian concept. This is
suggested by the authors' references, when discussing surplus measures, to
such things as gross revenues versus production costs, impacts in related
markets, and substitute and complementary goods. The difficulty may also
lie with the model's failure to distinguish between a short-run and a long-
run supply function. While the additional capital investment posited in
the analysis would shift the short-run supply function, such investment

would result only in movement <u>along</u> the stable long-run function. The simple procedure of comparing the areas above the short-run supply curve could be mistakenly interpreted as additional producers' surplus.

This relates to my final point which has to do with the general methodology of the paper and its relation to the elasticities employed. The methodology is comparative static, as depicted in figures 5 and 6. The temporal dimension is introduced by assuming or projecting growth rates for the various functions and taking comparative static "snapshots" in years 1990 and 2000. The elasticities employed are clearly short run, as exemplified by the discussion of Canadian lumber demand elasticity and its depiction as being perfectly inelastic in the aforementioned figures. To assume such a function and to then proceed to shift it at a predetermined rate is to build into the model a mechanism where nothing happens within the model--no U.S. import policy, no variation in the rate of growth of U.S. demand (which rate is not revealed in the paper, incidentally) no change in Canadian lumber production conditions--can have the slightest effect on Canadian consumption as of year X. My feeling is that the result is perhaps more of an arithmetical exercise than an economic one. I would be much more at ease with a model which used long-run elasticities to examine an essentially long-run phenomenon.

U.S. FOREST PRODUCTS TRADE AND THE MULTILATERAL TRADE NEGOTIATIONS

Samuel J. Radcliffe

The recent completion of the Multilateral Trade Negotiations (MTN) finished the process which was begun in 1973, when ministers from nearly one hundred countries signed the Tokyo Declaration, a statement of itent for the sixth round of trade negotiations held under the auspices of the General Agreement on Tariffs and Trade (GATT).

In the five rounds of negotiations held prior to the Tokyo Round, efforts had been concentrated on the reduction and elimination of tariff barriers to trade. The Kennedy Round of 1963-67 was the most successful in achieving that objective, as nonagricultural tariffs were reduced by about 35 percent (Preeg, 1970). Tariff reduction was also a goal of the Tokyo Round participants, but in addition, nontariff barriers were to be formally and extensively addressed in the negotiations. While it was sought to reduce or eliminate the trade-distorting effects of specific nontariff barriers, a potentially more important aim was to establish a framework under which classes of nontariff barriers could be brought "under more effective international discipline" (GATT, 1974).

The agreements made by the United States in the MTN are expected to have varying effects across the different sectors of the U.S. economy.

he purpose of this paper is to provide an assessment of the trade effects
hich might be expected in the forest products, that is, wood, wood pro-
ucts, and paper and paperboard sectors.

It is not our intent to comprehensively review the agreement's impact
n every wood-based commodity. Rather, our approach will be to focus on
he two major sectors of industrial forest products: wood products and
ulp, paper and paperboard. We will, however, discuss specific commodi-
ies which by themselves constitute major components of the two forest pro-
ucts sectors. We have limited our discussion to U.S. trade with Canada,
apan, and the European Economic Community (EEC).

This paper is divided into four sections. In the first section of the
aper, we briefly review the U.S. total forest products trade from 1960
o 1977, and the commodity structure of U.S. forest product imports and
xports. In the next section, we present an analysis of the impacts on
S. trade caused by U.S. and foreign tariff liberalization. We then re-
iew the provisions of the trade agreements relating to nontariff barriers,
nd note their potential for application in the forest products sector.
e conclude with a discussion and summary of our results.

An Overview of U.S. Forest Products Trade

The United States has historically figured quite prominently in the
orld's trade of forest products. Table 1 shows the United States and
orld forest products trade at five-year intervals beginning in 1960.

World trade in forest products has increased substantially since
960 with the average annual growth in nominal value greater than 10
ercent, a rate significantly higher than inflation, for the period
960-1977.

Table 1. U.S. and World Forest Products Trade, 1960, 1965, 1970, 1975, 1977
(value in U.S. $1,000)

| Year | World exports | United States | | U.S. exports as percentage of total world exports | U.S. imports as percentage of total world exports |
		Exports	Imports		
1960	6,216,960	608,456	1,565,401	10	25
1965	8,062,546	777,380	1,833,974	10	23
1970	12,563,469	1,622,856	2,300,551	13	18
1975	26,136,681	3,495,375	4,020,308	13	15
1977	33,351,771	3,902,336	5,459,263	12	18

Source: Food and Agricultural Organization, Organization of the United Nations, Yearbook of Forest Products, 1977 Annual (Rome, FAO, 1979).

U.S. exports, as a percentage of world exports, only slightly increased during the 1960-77 period, while U.S. imports significantly declined as a percentage of world trade, from 25 percent in 1960 to 18 percent in 1977. Nevertheless, the United States was the world's largest importer and the second-largest exporter of forest products in 1977.

The U.S. net deficit in forest products trade, relative to gross forest product imports, has substantially improved since 1960, when net imports represented 61 percent of gross imports. By 1977, that figure had declined to 29 percent.

The structure of the U.S. 1977 forest products imports and exports can be observed in tables 2 and 3. Canada was far and away the most important source of U.S. imports; the $5.4 billion worth of Canadian forest products represented 87 percent of all U.S. forest products imports. Japan and the European Economic Community (EEC) each captured only about 1 percent of total U.S. forest product imports, but were important suppliers of particular commodities. The remaining U.S. imports originated with a variety of supplying countries.

Just three commodities--softwood lumber, wood pulp, and newsprint-- constituted more than 80 percent of the value of U.S. forest product imports.

Shipments to Canada, the EEC, and Japan together comprised 71 percent (by value) of all U.S. forest product exports. The principal export commodities were softwood logs, wood pulp, and paper and paperboard. The three commodities contributed about 75 percent of the total 1977 export value of $4.1 billion.

Table 2. U.S. Imports of Forest Products by Area of Origin, 1977

(value in U.S. $1,000)

Commodity			Origin				Percentage of total forest products
Code	Description	Canada	EEC	Japan	Other	Total	
242.1	Pulpwood	10,803	--[a]	--	740	11,544	0
631.8EX	Pulpwood chips	28,579	--	--	2	28,581	0
242.2	Softwood logs	21,933	--	--	--	21,933	0
242.3	Hardwood logs	1,634	38	38	1,160	2,870	0
243.2	Softwood lumber	1,961,437	321	74	10,589	1,972,422	32
243.3	Hardwood lumber	30,735	764	1,007	91,675	124,180	2
631.1	Wood veneer	53,657	4,661	761	33,893	92,972	1
631.2	Plywood	13,994	905	75,708	359,648	450,255	7
631.4EX	Particle board	35,105	--	--	363	35,468	1
Total solid wood		2,157,877	6,689	77,588	498,070	2,740,225	44
251	Wood pulp and waste paper	1,194,327	1,299	1,524	42,420	1,239,571	20
641.1	Newsprint	1,879,056	--	1	16	1,879,073	30
641EX	Paper and paperboard less newsprint	218,004	52,146	13,158	103,064	386,373	6
Total pulp, paper and paperboard		3,291,387	53,445	14,683	145,500	3,505,017	56
Total Forest Products		5,449,264	60,134	92,271	643,570	6,245,242	100

Source: U.S. Bureau of the Census, U.S. General Imports/Schedule A Commodity Groupings by World Area, Report FT-150, Annual 1977 (Washington, D.C., GPO, 1978).

[a] Dashes denote zero value.

140

Table 3.   U.S. Exports of Forest Products by Area of Destination, 1977

(value in U.S. $1,000)

Commodity		Destination					Percentage of total forest products
Code	Description	Canada	EEC	Japan	Other	Total	
242.1	Pulpwood	8,213	1	--[a]	--	8,214	0
631.8EX	Pulpwood chips	2,501	3	168,178	7,541	178,223	4
242.2	Softwood logs	35,111	892	810,973	51,771	898,747	22
242.3	Hardwood logs	7,648	43,220	7,449	14,082	72,399	2
243.2	Softwood lumber	103,983	116,345	104,584	119,484	444,396	11
243.3	Hardwood lumber	56,990	33,707	2,308	10,376	103,381	3
631.1	Wood veneer	10,175	27,551	215	8,801	46,742	1
631.2	Plywood	16,404	44,102	1,090	14,161	75,757	2
631.4EX	Particle board	7,810	48	353	4,352	12,563	0
Total solid wood		248,835	265,869	1,095,150	230,568	1,840,422	45
251	Wood pulp and waste paper	58,595	436,296	181,072	371,611	1,047,574	26
641.1	Newsprint	1,922	6,152	145	34,327	42,545	1
641EX	Paper and paperboard, less newsprint	265,675	269,200	61,130	523,854	1,119,859	27
Total pulp, paper, and paperboard		326,192	711,648	242,347	929,792	2,209,978	55
Total Forest Products		575,027	977,517	1,337,497	1,160,360	4,050,400	100

Source: U.S. Bureau of the Census, U.S. Exports/Schedule B Commodity Groupings by World Area, Report FT-450, Annual 1977 (Washington, D.C. GPO, 1978).

[a] Dashes denote zero value.

Tariff Concessions in the Tokyo Round

The tariff concessions agreed upon by the Tokyo Round participants are expected to directly effect reductions in the prices of imported goods and subsequent increases in the quantity of those imports. In the following analysis we make rough estimates of the magnitude of increases in U.S. forest products trade that might be expected as a result of the tariff negotiations.

Methodology

Our estimates are based on the standard comparative statics model of the trade effects of a tariff reduction (or imposition). This approach is fairly simple, but a review of the fundamentals of the model may facilitate interpretation of the results.

Figure 1 depicts a country's import demand and supply curves DD and S'S' in the presence of a tariff of P_1P_2 per unit. The equilibrium quantity is Q_1, the price to domestic consumers is P_1, and the world price is P_2. If the tariff is removed, supply shifts downward to SS, and the new equilibrium quantity is Q_2 at the world price P_2.

The estimates which are presented in the next section are of the increase in the value of the country's imports, represented by rectangle Q_1ABQ_2. The arithmetic involved in quite simple. For each country, at the tariff line level, the percentage change in price due to the reduction in tariff is calculated.[1] The percentage price change is then applied to

---

[1] The percentage change in price ΔP, is calculated: $\Delta P = \dfrac{(t_1 - t_2)}{1 + t_1}$ where t_1 is the initial tariff and t_2 is the reduced tariff (ΔP, t_1, and t_2 in decimal forms).

Figure 1. Import demand and supply, with and
without a tariff

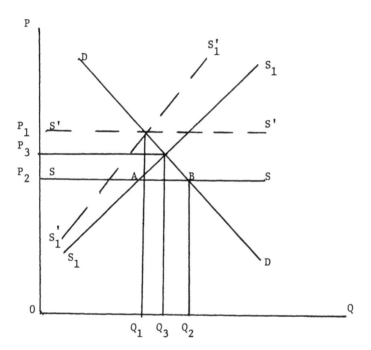

the relevant import demand elasticity. The resulting figure is not only

the percentage increase in quantity of imports, but also the percentage

increase in the value of imports. This equality would not generally hol[...]

but does so under the assumption of infinite supply elasticity.[2] The pe[...]

centage increase in value is then applied to the 1977 value of imports o[...]

---

[2]Referring to figure 1, for the infinitely elastic curves SS and S'[...]
the percentage increase in the quantity of imports, ΔQ, is equal to:

$$\Delta Q = \frac{Q_2 - Q_1}{Q_1}$$

and the percentage increase in the value of imports, ΔV, is equal to:

$$\Delta V = \frac{(Q_2 - Q_1)P_2}{Q_1 P_2} = \frac{Q_2 - Q_1}{Q_1}$$

therefore $\Delta V = \Delta Q$ for the infinitely elastic supply curves. But for cur[...]
which are less than infinitely elastic, such as $S_1 S_1$ and $S'_1 S'_1$,

$$\Delta Q = \frac{Q_3 - Q_1}{Q_1}$$

$$\Delta V = \frac{\left[(Q_3 - Q_1)P_3\right] + \left[(P_3 - P_2)Q_1\right]}{Q_1 P_2} = \frac{(Q_3 - Q_1)P_3}{Q_1 P_2} + \frac{(P_3 - P_2)}{P_2}$$

Since $P_3 > P_2$, $\Delta V > \Delta Q$ for the less than infinitely elastic supply curve[...]

the particular tariff line item to arrive at the absolute increase in import value.[3]

The assumption of infinite supply elasticities, although convenient, is not arbitrary. Trade studies have often incorporated this assumption not only because there is a lack of detailed estimates of supply elasticities but also because the empirical estimates of export supply elasticities as do exist are indeed very high or infinite (Cline and coauthors, 1978). Thus, the assumption of infinite supply elasticities probably does not damage our results, and simplifies the algebra involved in the estimation.

Moreover, the assumption of infinite supply elasticities provides an "upper bound" estimate of the increase in the physical volume of imports resulting from a tariff reduction. This feature is useful for estimating the maximum impact of a tariff reduction upon physical production levels as well as trade flows.

Published estimates of import demand elasticities for individual forest products or for relatively narrowly defined product groups were not avail-

---

[3] Sources of import data are:

United States - U.S. Bureau of the Census, U.S. Imports for Consumption and General Imports, Report FT-246, Annual 1977 (Washington, D.C., USGPO, 1978).

Canada - Unpublished materials from statistics Canada, Ottawa (1974 import data).

Japan - Customs Bureau of Ministry, Finance of Japan, Japan Exports and Imports Country by Commodity, January-December 1977 (Tokyo, Japan Tariff Association, 1978).

European Community - Statistical Office of the European Communities, Analytical Tables of Foreign Trade NIMEXE 1977, vol. E (44-49) (Belgium, 1979).

able for this study. However, a Brookings Institution study (Cline and
and coauthors, 1978, p. 58) of trade liberalization does report, for the
United States, Canada, Japan, and the EEC, import demand elasticities for
the broad product categories "Wood and Articles of Wood" and "Paper and
Paperboard Articles" (table 4). Our use of these elasticities for the
individual tariff line calculations means that results at the tariff item
level are unlikely to be reliable, but there is no reason to believe that
the sum of the tariff line results is a biased estimate of the increased
value of imports for the broad product sectors.

The results which are reported below for the wood and paper sectors
were obtained in essentially the following manner. After making estimates
of the increase in import value due to tariff liberalization for each
tariff item, the individual increases were aggregated to arrive at an increase for the sector as a whole. The percentage increase which resulted
for the entire sector was then applied to the 1977 U.S. import or export
value for the relevant sector to obtain the value of increased U.S. imports of exports in 1977 U.S. dollars.[4]

An important feature of the analysis is that the increases in U.S.
imports and exports resulting from the tariff neogitations are estimated
for each country or area of origin and destination. In order to accomplish
this, it was assumed that, for U.S. exports, the United States will retain
its 1977 share of the foreign country's total imports of each commodity.

---

[4]This method simplified the conversion of foreign currencies to U.S.
dollars, and the updating of the 1974 Canadian import data to 1977. Moreover, it eliminated any problems which might have arisen in the conversion
of c.i.f. values to f.o.b. values.

Table 4. Import Demand Elasticities Employed in the Estimation of In-
          creased Imports for the United States, Canada, Japan, and the
          European Community

Country	Solid wood commodities	Pulp, paper, and paperboard
United States	-0.96	-1.44
Canada	-2.14	-2.07
Japan	-1.33	-1.74
EEC	-0.83	-0.79

Source: William R. Cline, Noboru Kawanabe, T. O. M. Kransjo, and
Thomas Williams, Trade Negotiations in the Tokyo Round: A Quantitative
Assessment (Washington, D.C., Brookings Institution, 1978) p. 58.

For example, if the United States supplied 25 percent of the European
Community's total kraft linerboard imports in 1977, then 25 percent of
the EEC's total increase in linerboard imports is attributed to an increase
in U.S. exports to the EEC. In a like manner, foreign countries are assumed to retain their 1977 shares of total U.S. imports of each commodity, and
thus the total increase in U.S. imports can be distributed among the
various supplying regions.

The estimates of increased import values at the tariff line level are
not totally without value. The elasticities used for the product groups
can be considered, at worst, first-approximations of the true elasticities
for the individual commodities which constitute the group.[5] Thus, the
results for individual commodities do provide a rough indication of which
producers are likely to become significant contributors to the overall
trade increases predicted for the major groups of forest products.

Table 5. Pre—Tokyo Round and Concession Tariff Rates on Selected U.S. Forest Product Imports

TSUSA Item	Description	Tariff rate	
		Pre—Tokyo	Concession[a]
202.0340	Spruce lumber, dressed or worked	Free	Free
202.0940	Pine lumber, dressed, except eastern white, red or parana pine	Free	Free
202.4020— 202.4040	Lauan lumber, rough or dressed	$.75/MBF (0.2%)	Free
240.0020	Birch veneers, not reinforced or backed	4%	Free
240.0200	Lauan veneers, not reinforced or backed	10%	Free
240.0320	Hardwood veneers, except birch, maple, or lauan	5%	Free
240.0340	Softwood veneers	5%	Free
240.1740	Lauan plywood, not face finished	20%	8%
240.2360	Hardwood plywood, not face finished, except birch, lauan, walnuts, sen, or mahogany	10%	8%
240.2100	Softwood plywood, not face finished, except spanish cedar and european red pine	20%	8%[b]

245.0000–245.2000	Hardboard, not face finished	7.5%	3%
245.5000	Wood particle board	10%	4%
250.0280	Bleached sulphate softwood pulp	Free	Free
252.6500	Standard newsprint paper	Free	Free
252.6720–252.6740	Uncoated book and printing paper	$.0008/LB + 2% (2.5%)	Free
254.4600	Coated or impregnated printing paper, not lithographically printed, except India or Bible paper	$0.1/LB + 2% (6.4%)	2.5%
256.0500	Wallpaper	5%	Free

Note: Equivalent ad valorem rates are given in parenthesis.

Source: General Agreement on Tariffs and Trade, Geneva (1979) Protocol to the General Agreement on Tariffs and Trade, Schedule XX--United States, and U.S. International Trade Commission, Tariff Schedules of the United States Annotated (1976) ITC Publication 749, (Washington, D.C., GPO, 1975).

[a]Full concession rate shown will become effective on January 1, 1987, following staged reductions beginning January 1, 1980. Full concession rate on 202.40, lumber, will become effective January 1, 1980; on 240.00, 240.02, and 240.03, veneers--January 1, 1981; on 240.17, plywood--January 1, 1984; and on 252.67, paper--January 1, 1980.

[b]Contingent upon establishment of a North American plywood standard.

Results for U.S. Imports

Pre-MTN and concession tariff rates on some important U.S. forest product imports are shown in table 5. A comparison of the tariff rates with
the U.S. 1977 import values shown in table 2 reveals that more than 80
percent of U.S. forest product imports were duty-free before the negotiations; thus, any increase in imports resulting from the MTN will necessarily be small in relation to total U.S. forest product imports.

For the commodities which had been subject to duty prior to the negotiations, some significant concessions were made. Veneer and plywood had
and will still, face the stiffest tariffs levied on U.S. forest product
imports. However, under the new tariff schedule, more than three-fourths
of all veneer imports will be duty-free; in addition, tariffs on most of
the important U.S. plywood imports will be harmonized at 8 percent, elimi-
nating the tariff disparity which formerly existed between lauan plywood
and other tropical hardwood plywoods. A potentially important reduction
the tariff on softwood plywood, from 20 percent to 8 percent, will not be
implemented until Canada and the United States can agree on a North American plywood standard. Currently, much of the low-grade plywood which is
used extensively in the United States for sheathing and other nonvisible
applications is not accepted in Canada, in either its plywood standards o-
its building codes. Negotiations on a North American standard, however,
are reported to be at a stalemate.

---

[5]The notion that all of the products under consideration could have
equal demand elasticities is not ruled out in McKillop's study of U.S.
forest products demand and supply. His interval estimates of demand elasticities for six different forest products and product groups overlap in
all but two of fifteen possible pairings (McKillop, 1967).

Table 6. U.S. Imports and Estimated Increases in Imports of Major Forest
Product Sectors, by Area of Origin

	Solid Wood Commodities			Pulp, Paper, and Paperboard		
	1977 Import value ($1000)	Increase in import value ($1000)	(%)	1977 Import value ($1000)	Increase in import value ($1000)	(%)
Canada	2,157,877	4,997	0.2	3,291,387	8,444	0.3
EEC	6,689	63	0.9	53,445	2,515	4.7
Japan	77,588	3,331	4.3	14,683	670	4.6
Others	498,070	27,210	5.5	145,500	5,245	3.6
Total	2,740,225	35,601	1.3	3,505,015	16,874	0.5

Table 6 presents our estimates of the increases, due to tariff liberalization, in the value of U.S. imports in the wood and pulp, paper and
paperboard sectors.

As mentioned above, it has been assumed that every foreign country
will continue to supply its 1977 share of U.S. imports of each commodity.
The percentage increases in U.S. forest product sector imports, shown in
table 6, differ by country because in 1977 every country did not export the
same commodity bundle to the United States.

More than half of the total increase in forest product imports will
be in solid wood commodities from countries other than Canada, Japan, or
the European Community. Most of this increase will take the form of hard-

wood plywood from Korea and Taiwan. The increase in imports of solid wood products from Japan is also largely hardwood plywood.

Canada's increase in exports of solid wood products to the United States will principally in be birch veneers and in particle board.

In the paper and paperboard group, U.S. increased imports represent only a small percentage of the total imports in the sector; as mentioned above, this results because wood pulp and newsprint, which were duty-free before the MTN, make up such a large fraction of total U.S. paper and paperboard imports (89 percent by value in 1977). If pulp and newsprint are excluded from the total, the sector's imports increase by 4.4 percent.

The important increases in paper and paperboard imports from Canada will be in uncoated and coated printing papers, and kraft paper and paperboard; on most of these commodities tariffs which had ranged in the neighborhood of 2.5 to 4 percent have been eliminated.

Imports of coated printing papers from Finland and West Germany will also expand significantly, as will hardboard from Brazil and wallpaper from the European Community.

To gain another perspective on the significance of the estimated increases in U.S. imports of hardwood veneer and plywood and paper and paperboard, it will be useful to relate the estimates to the U.S. apparent consumption of these commodities. In 1976, imports of hardwood veneer and plywood represented about 70 percent of U.S. consumption (Phelps, 1978). Thus, the predicted 4 to 6 percent increases in imports will mean a 3 to 4 percent increase in U.S. consumption of hardwood plywood and veneer. For paper and paperboard, expanded imports will impact very little on U.S. consumption. Imports of paper and paperboard other than newsprint are only 1 to 2 percent of apparent consumption (U.S. Department of Commerce, 1977), so that the predicted import gains of 4 to 5 percent will effect a negligible increase in U.S. consumption.

Results for U.S. Exports

As table 2 indicates, a large percentage of the U.S. forest product exports consists of softwood logs, softwood lumber, and wood pulp. These commodities were generally not subject to foreign tariffs before the MTN; the expansion of U.S. exports due to tariff liberalization will therefore be small in relation to total U.S. forest product exports, although significant gains may be made in the export of specific commodities.

Tables 7, 8, and 9 show the tariff concessions made by Canada, Japan, and the European Community on selected forest products. The "concession rate" column indicates the final tariffs to be levied on goods; the tariffs will be implemented in stages over eight to ten year periods beginning January 1, 1980.[6]

---

[6]The staging schedules vary by commodity and country. For simplicity, we have assumed immediate reductions to the full concession rates. The present value of the staged import/export increases will, of course, be smaller than our estimates for any interest rate greater than zero. A numerical example may help to illustrate the magnitudes involved. Assuming an import demand elasticity of -1.74 on Japanese imports of wood pulp, the increase in U.S. exports resulting from an immediate reduction in the Japanese tariff (from 5 percent to 2.2 percent) would amount to $8.4 million, or 4.6 percent of 1977 U.S. exports. The present value (using a 10 percent interest rate) resulting from an equivalent tariff reduction staged in eight equal annual cuts of 0.35 percent each is $6.3 million, which is 3.5 percent of 1977 exports, and 25 percent less than the gains resulting from a one-time reduction. Our estimates are thus higher than a more realistic analysis would indicate, but the relative magnitudes are probably not seriously biased, since the staging schedules for most commodities are very similar.

Table 7. Pre-Tokyo Round and Concession Tariff Rates on Selected Canadian Forest Product Imports

Tariff item	Description	Tariff rates	
		Pre-Tokyo	Concession[a]
5000-1	Logs, poles, posts, pilings, railway ties, etc.	Free	Free
50040-1- 50045-1	Timber or lumber of any species, not further manufactured than sawn and planed	Free	Free
50600-9	Softwood plywood, face finished	15%	8%[b]
50705-1	Single ply veneers of wood, not taped or jointed	7.5%	Free
50715-4	Softwood plywood; plywood with a face veneer of softwood	15%	8%[b]
2000-1	Pulp of wood, straw, or any other vegetable material	Free	Free
19200-3	Hardboard, unfinished	15%	6.5%
19500-1	Hanging paper or wallpaper	15%	7.5%
19700-3	Uncoated paperboard	15%	9.2%
19700-4	Corrugating medium, not cut to size or shape	15%	4%
19750-3	Coated groundwood printing paper	12.5%	2.5%
19750-4	Uncoated groundwood printing paper	12.5%	Free

Source: GATT, Geneva (1979) Protocol, Schedule V--Canada, and Revenue Canada, Customs and Excise, Customs Tariff and Amendments, (Ottawa, 1978).

[a] Full concession rate shown to become effective January 1, 1987, following staged reductions beginning January 1, 1980. The full concession rate on 50710-1, veneers, will become effective January 1, 1982.

[b] Contingent upon establishment of a North American plywood standard.

Table 8. Pre-Tokyo Round and Concession Tariff Rates on Selected Japanese Forest Product Imports

CN ode	Description	Tariff rates	
		Pre-Tokyo	Concession[a]
.03-326	Sawlogs and veneer logs of hemlock	Free	Free
.05-512	Sitka spruce lumber	Free	Free
.09-211	Softwood pulpwood chips	Free	Free
.14-230	Sheets for plywood	15%	15%
.15-191	Plywood with both faces of softwood	15%	15%
.18-100	Reconstituted wood in sheets or boards	20%	12%
.01-121	Sulphite wood pulp, dissolving grades	5%	2.2%
.01-126	Bleached or semibleached kraft wood pulp, except dissolving grades	5%	2.2%
.02	Waste and scrap paper	Free	Free
.01-310	Kraft liner, weighing more than 30 g/m^2 but not more than 300 g/m^2	12%	7%
.01-321 .01-329	Kraft paper	12%	10%
.01-410 .01-420	Kraft liner and kraft paperboard weighing more than 300 g/m^2	8%	5%
.07-991	Paper and paperboard, coated or impregnated with artificial resins	8%	5.1%

Source: GATT, Geneva (1979) Protocol, Schedule XXXVIII--Japan, and pan Tariff Association, Customs Tariff Schedules of Japan, (Tokyo, 1977).

[a]Full concession rate shown to become effective January 1, 1987, llowing staged reductions beginning January 1, 1980. The full concesion rates shown for 48.01-310, 48.01-321-329, kraft paper and liner, ll not become effective until January 1, 1992, and for 48.01-410-420, aft linerboard, will not become effective until January 1, 1990.

Table 9. Pre-Tokyo Round and Concession Tariff Rates on Selected Euro-
          pean Community Forest Product Imports

CCT code	Description	Tariff Rates	
		Pre-Tokyo	Concession[a]
44.03B	Wood in the rough, except poles	Free	Free
44.05A, C	Wood sawn lengthwise, sliced or peeled, but not further prepared of a thickness exceeding 5 mm	Free	Free
44.14B	Veener sheets and sheets for plywood	7%	6%
44.15	Softwood plywood, of a thickness greater than 8.5 mm, or, if sanded, of a thickness greater than 18.5 mm	13%	10%[b]
44.15	Other plywood, blockboard, laminboard etc.	13%	10%
47.01	Pulp derived by mechanical or chemical means from any fiborous vegetable material	Free	Free
48.01C	Kraft paper and kraft board, except for the manufacture of paper yarn or large-capacity sacks	8%	6%
48.01F	Paper and paperboard in rolls or sheets except newsprint, cigarette paper, kraft paper and board, paper for stencil making, and hand-made paper and board	12%	9%
48.07C	Coated paper and paperboard, except ruled, lined, or squared, coated with mica, or bleached and coated with artificial plastic material	12%	9%

Source: GATT, Geneva (1979) Protocol, Schedule LXXII--European Community and Official Journal of the European Communities, vol. 20, no. 1289 (November 14, 1977).

[a]Full concession rates shown to become effective January 1, 1987, following staged reductions beginning January 1, 1980.

[b]Annual tariff-exempt quota of 400,000 m^3 of softwood plywood was raised to 600,000 m^3.

Canada has made the largest tariff reductions of the three areas, especially on paper and paperboard products. Nearly 60 percent of the total estimated increase in U.S. exports (table 10) is in exports of paper and paperboard to Canada. In particular, fiberboard, strawboard, and other paperboard and coated and uncoated printing papers will be exported from the United States to Canada in significantly greater quantities.

Increased Canadian imports of solid wood commodities will consist largely of particleboard and wood veneer. Because the reduction in the Canadian tariff on softwood plywood depends on the progression of negotiations towards a common North American standard, the rise which might result from such a reduction was not included in our estimate of expanded U.S. exports. The potential impact of an agreement on the standards issue is not insignificant; the resulting tariff reduction would more than double our estimate (table 10) of the gain in U.S. solid wood exports to Canada.

Japan agreed to very few significant tariff reductions on solid wood commodities. The tariff on particleboard will be lowered from 20 percent to 12 percent, but veneer and softwood plywood will still be subject to 15 percent _ad valorem_ duties.

Unlike most other industrialized nations, Japan charges a duty on imports of wood pulp, but the reduction in that duty (from 5 percent to 2.2 percent) will enable the United States to expand its already large pulp exports to Japan. Other important components of Japan's increased paper and paperboard imports include kraft liner and paperboard, and coated paper and paperboard.

Table 10. U.S. Exports and Estimated Increases in Exports of Major Forest Product Sectors, by Area of Destination

	Solid wood commodities			Pulp, paper, and paperboard		
	1977 export value ($1000)	Increase in export value ($1000)	(%)	1977 export value ($1000)	Increase in export value ($1000)	(%)
Canada	248,335	5,504	2.2	326,192	31,842	9.8
EEC	265,869	329	0.1	711,648	5,210	0.7
Japan	1,095,150	84	0.01	242,347	10,575	4.4
Others	230,568	a	–	929,792	a	–
Total	1,840,422	5,917	0.3	2,209,979	47,609	2.2

aNot estimated, but probably small (see text).

The European Community will experience only small percentage increases in its forest products imports from the United States, as nearly two-thirds of the U.S. exports to the EEC entered duty-free prior to the negotiations.

In the solid wood products sector, plywood and veneer are important U.S. exports subject to the EEC's external tariff. However, the EEC had allowed a tariff-exempt quota of 400,000 cubic meters of softwood plywood and, as a result of the negotiations, expanded the quota to 600,000 cubic meters. These quotas are the officially bound rates; in practice, the EEC has for a number of years applied larger quotas which allow virtually all softwood plywood duty-free entry. The EEC's tariff reduction on softwood plywood will thus have no effect on actual U.S. exports.

The predicted increase in U.S. exports of paper and paperboard to the European Community (table 10) will consist largely of coated paper and paperboard, and kraft paper and board. While the reductions in the EEC's external tariff on paper and paperboard represent an improvement over the existing tariff situation, they do not eliminate the disparity between tariffs paid by the United States and tariffs paid by Sweden and Finland on goods imported by the EEC. The Nordic countries enjoy favorable treatment from the European Community as a result of the Free Trade Agreements negotiated between the EEC and members of the European Free Trade Association (EFTA) when the European Community was enlarged in 1973. Finland and Sweden, which are the largest suppliers of paper and paperboard to western Europe, have been benefiting from EEC tariff reductions since 1974. These reductions are scheduled to progress until 1984, at which time the important paper and paperboard commodities will be duty-free. The United

States will still be facing tariffs of 6 to 9 percent when the final stages of reductions in the Common External Tariff is implemented in 1987.

Estimates of the gain in U.S. exports to countries other than Canada, Japan, and the EEC have not been made in this paper. But by observing the commodity structure of U.S. exports to these "other" countries (table 2), potential export expansion opportunities can be identified.

Nearly half the U.S. forest products exports are in goods which generally are relatively unprotected by tariffs--unprocessed wood, lumber, and wood pulp. Here we might expect little or no tariff liberalization, and hence insignificant export gains.

Most of the remaining U.S. exports are paper and, in particular, paperboard commodities. The major importers in this group are developing nations, which would be expected to offer few concessions in multilateral negotiations involving the industrialized countries (Cline and coauthors, 1978, p. 208). Again, there would seem to be few opportunities for significant U.S. export expansion to countries other than Japan, Canada, and the European Community.

Nontariff Barriers

The negotiation of and agreement to international codes relating to classes of nontariff barriers was a major feature which distinguished the Tokyo Round from all previous multilateral trade negotiations. There were, in all, ten agreements signed by the United States in Geneva. Three agreements which could have important impacts on U.S. forest products trade are the codes relating to subsidies and countervailing measures, technical standards, and import licensing. Because of the nature of the barriers,

t is difficult to quantitatively estimate their impact on trade flows,
nd we have not attempted such estimation in this paper. Rather, we will
ry to indicate the extent to which these nontariff barriers are prevalent
n forest products trade, and explain how the new agreements might reduce
heir trade-distorting effects.

ubsidies/Countervailing Measures

Government subsidization of an industry allows that industry to oper-
te at lower private cost than would otherwise be the case. Although the
rimary objectives of such policies is not always related to international
rade, an artificial competitive advantage for the domestic producers over
oreign producers is created, thus simultaneously promoting exports and
iscouraging imports.

The pulp and paper industry have been heavily subsidized by govern-
ents throughout the world. Canada and some western European nations, im-
ortant to the United States both as a source of supply and as outlets for
merican production, have quite aggressively subsidized their domestic in-
ustries in order to enhance employment and economic development opportuni-
ies.

Subsidization has not been confined to the pulp and paper sector.
oth Korea and Taiwan, which together provided nearly 70 percent of the
.S. imports of plywood in 1977, subsidize their domestic plywood producers.

Subsidization schemes can be grouped into two categories (1) "domestic
ubsidies" which are intended primarily to correct problems in a country's
omestic economy, but which nevertheless may have international spillover
ffects; and (2) "export subsidies" which are aimed directly at increasing
e amount of a country's exports.

Under the new agreement, export subsidies are prohibited, unless granted to producers of certain primary products. Some of the relatively unprocessed forest products are included in the GATT definition of primary products, but most of the countries which have a substantial timber resource tend to restrict, not subsidize, exports of primary forest products

The agreement recognizes that the domestic aims of subsidies may be important, especially for developing countries, but that the trade-distort ing effects of such policies may be large, and should be avoided. In addi tion, the new code provides that if a domestic or third-country industry suffers material injury, or the threat of injury, from the import of sub sidized goods, then the importing country may levy countervailing duties to offset the trade effects of such subsidies.

Included in the code is a procedural framework for notification of subsidies, determination of injury, dispute settlement, and GATT authori zation to use countermeasures. The implementation of these procedures could go a long way toward either the reduction or elimination of the trade-distorting effects of subsidies, or the promotion of the use of cour tervailing duties.

Technical Standards

Disparities in standards with respect to engineering strengths, dimer sions, quality, sanitary conditions, and so forth are believed to be a major nontariff barrier which restricts both countries exporting to the United States and the United States exporting to foreign countries. That is, when two countries have different standards for structural lumber, for example, then a producer of lumber must be able to adapt to two different production processes, one for the domestic market and one for the export

market. This involves increased costs which may make the export of lumber prohibitive.

Inspection, testing, grading, and labeling requirements of the importing country also involve expense to the suppliers over and above the normal
costs of production and transportation. For example, it has been estimated
that the examination and sampling procedure for lumber imported into Japan
adds $12 to $16 per thousand board-feet to the cost of the lumber (Forster,
1978).

Standards for wood products are often dictated by the construction
techniques used in a country, which in turn reflect the building codes in
existence. For most products, this does not seem to be a significant
barrier to trade between the United States and Canada because species,
technologies, and end products in those two countries are essentially identical. The U.S.-Canadian disagreement over plywood standards is an important exception. American and Canadian exporters of processed timber
have, however, reportedly faced serious problems due to the application of
dimension and quality standards in Japan and certain European countries.
The MTN did not prove to be a suitable forum for discussion of specific
standards-related barriers; most of these will have to be addressed bi-
laterally.

The agreement on technical standards does, however, list several
guidelines which countries are expected to follow in formulating product
standards. The most important of these, for wood products, stipulates
that, where possible, standards should be based on performance rather than
design characteristics. In addition to the guidelines, the agreement pro-

vides a procedure for the resolution of dispute involving standards and technical regulations.

Import Licensing

Import licensing is often used as a means of collecting statistical data or administering quotas or other regulations, but its most controversial use in the field of trade barriers is a method of government control of trade. The amount of trade to be permitted is often not made public, and delays in processing applications, along with discriminatory granting of licenses, are not uncommon. Because of the uncertainty created, traders have complained of licensing practices more than of any other single type of trade barrier. The U.S. Tariff Commission (1974) reports that more than one-fourth of the complaints which the Commission receives pertain to licensing practices.

For forest products, most of the countries which have licensing requirements are developing nations. Import licenses are required for paper and paperboard products more frequently than for any other wood-based commodity (U.S. Department of Commerce, n.d.).

The Multilateral Trade Negotiations agreement in import licensing seeks to make licensing requirements and procedures clear and neutral, and provides a means for settlement of disputes.

Discussion and Summary

Economists have identified a number of impacts which result from changes in a country's amount and/or structure of trade protection. Traditionally, concern has been expressed over the trade, employment, and welfare effects of such changes. In this paper, we have estimated the impacts on U.S. forest products trade of the tariff reductions which were agreed upon in the recent Multilateral Trade Negotiations. While the estimation of welfare and employment effects might seem to be a logical next step, we have not carried our analysis further; we will, however, make some comments below on the implications of the predicted trade increases for U.S. forest industry employment.

A fourth occurrence attributed to changes in the structure of protection is a reallocation of resources among the various industries within an economy. The appropriate concept for analysis in this case is the effective protective rate, which incorporates both the tariffs on final consumption goods and the tariffs on inputs to the production of those goods. The use of effective protective rates in analysis is beset with theoretical as well as practical difficulties.[7] Furthermore, data which are generally available permit the calculation of effective protection for only fairly broad sectors of the economy. Here detailed intput-output studies of the forest products sectors would be of tremendous value. We have not considered effective protective rates in this paper, but the subject is an important and potentially fruitful area for research on the forest products industry.

The results of our analysis of the trade increases from tariff liberalization indicate only minor impacts on the major forest product sectors, because most commodities of importance in U.S. trade, both imports and exports, received relatively little nominal protection before the MTN.

The largest increases expected for U.S. forest product imports will be made in shipments of hardwood plywood from Korea, Taiwan, and Japan. The United States will also experience increased shipments of printing

---

[7]For a general discussion of effective protection, see (Grubel, 1971).

paper from Canada, Finland and Sweden, but these increases represent small fractions of total U.S. printing paper imports.

For U.S. exports, relatively large gains might be made in paper and paperboard shipments to Canada, but the predicted percentage gain in total U.S. paper and paperboard exports is only about 2 percent. Despite tariff reductions by the European Community, the United States will still be facing significantly higher tariffs than will Finland and Sweden in the EEC paper and paperboard market.

U.S. solid wood product exports are predicted to increase by less than half of one percent. What appears to be a significant liberalization of the European Community's quota and tariff on softwood plywood will in fact have no effect on U.S. exports because of the EEC's application of duty-free quotas which exceed their bound rates. A potentially important reduction in the Canadian duty on softwood plywood awaits the negotiation of a common North American plywood standard.

A further perspective on the expected trade increases and their implications for direct U.S. employment increases can be gained by comparing U.S. forest products trade and production figures. For all forest products, in 1976, imports were 24 percent and exports were 13 percent of domestic production (Phelps, 1978, p. 39). Our predicted increases of 0.84 percent in total imports and 1.32 percent in total exports would lead to, at most, a production decrease of 0.2 percent due to increased imports, and a production increase of 0.17 percent due to export expansion. The net effect is thus a negligible 0.03 percent decline in production. This is a very rough calculation of aggregate production effects; calculations at a more detailed product level, however (with the possible exception of

rdwood plywood) would predict production effects of a similar magnitude.
der an assumption of fixed-factor proportions, the direct employment
pact on the forest products industry would also be negligible.

The impact on U.S. trade of the agreements on nontariff barriers is
tremely difficult to assess. The nature of the agreements, however,
ggests that any effects would have to be evaluated over a longer time
riod than would the effects of tariff liberalization, since the nontariff
reements, in general, provide frameworks for a more orderly assessment
d resolution of problems involving classes of nontariff barriers, rather
an the elimination of specific barriers.

168

REFERENCES

Cline, William R., Noboru Kawanabe, T. O. M. Kransjo, and Thomas Williams. 1979. Trade Negotiations in the Tokyo Round: A Quantitative Assessment (Washington, D.C., Brookings Institution).

Forster, Robert B. 1978. Japanese Forestry: The Resources, Industries, and Markets, Canadian Forestry Service, Information Report E-X-30 (Ottawa, Ontario, Department of the Environment).

General Agreement on Tariffs and Trade. 1974. GATT Activities in 1973 (Geneva, Switzerland, GATT).

Grubel, Herbert G. 1971. "Effective Tariff Protection: A Non-Specialist Introduction to the Theory, Policy Implications and Controversies," in Herbert G. Grubel and Harry G. Johnson, eds., Effective Tariff Protection (Geneva, Switzerland, General Agreement on Tariffs and Trade, Graduate Institute of International Studies).

McKillop, William. 1967. "Supply and Demand for Forest Products--An Econometric Study," Hilgardia vol. 38, no. 1 (March).

Phelps, Robert B. 1978. The Demand and Price Situation for Forest Produc 1976-77, U.S. Department of Agriculture, Forest Service, misc. public tion no. 1357 (Washington, D.C., GPO).

Preeg, Ernest H. 1970. Traders and Diplomats (Washington, D.C., Brookings Institution).

U.S. Department of Commerce. 1977. Pulp, Paper and Board vol. 33 no. 1 (Spring) (Washington, D.C., GPO).

_____. n.d. Unpublished materials.

U.S. Tariff Commission. 1974. Trade Barriers, TC Publication 665, Part 2 "Non-Tariff Trade Barriers," vol. 5 (Washington, D.C.).

Discussion by Harold Wisdom

I am in general agreement with Sam Radcliffe's estimates of the poten-
ial impact of the Multilateral Trade Negotiations (MTN) on international
rade in forest products. The fact that the pre-Tokyo Round tariff rates
n most forest products were relatively modest does suggest that the effect
f reductions in the already modest rates is likely also to be modest. My
eneral agreement with Radcliffe's estimates is, however, subject to sev-
ral reservations. These reservations are related to the use of the com-
arative-static, two-country model to analyze the effects of a general re-
uction in tariffs. A reduction involving many commodities and many
ations.

As Radcliffe explained, the estimates of the effect of the tariff re-
uctions were made by considering the impact upon each commodity group
eparately. First, the increase in imports of the commodity stimulated by
he tariff reduction was estimated. The increase was measued as the diffe-
ence in imports before the reduction and imports after the reduction, all
ther things remaining constant. That is, it is assumed that nothing
appens to affect either the position or the slopes of the import demand
nd supply functions.

The increase in imports was then allocated among the supplying coun-
ries on the basis of each country's share of total 1977 imports by the

importing country. The assumption is that the export supply of all existing suppliers is infinitely elastic. That is, the extra supply is assumed
to be available at constant cost. All existing suppliers will contribute
to the extra supply in exactly the same proportion as they have in the past

The problem with this approach is that the MTN is much more complex
than a simple, two-country trading situation. MTN involves general tariff
concessions on a wide range of commodities in many nations, with the concessions by individual nations ranging from negligible to substantial.
Thus, at the same time that the tariffs faced by country A on forest products are being reduced--improving country A's ability to compete in the
world market for those commodities, country A is also lowering its own
tariffs on forest products, making its domestic markets more vulnerable
to external competition. To further complicate matters, tariffs on competing and complementary, nonforest products also are being reduced simultaneously, introducing the possibility of complex substitution effects
and relative price changes. This problem situation surely strains the
analytic limits of the simple, two-country trade model.

Let me say, in passing, that I am fully aware that one can carry the
interdependence argument to absurd lengths, to the point where everything
depends on everything else. I certainly do not wish to go to that length.
However, we cannot ignore the fact that the MTN does imply an extremely
complex set of economic adjustments; adjustments that are bound to occur
and that cannot be assumed away by adopting the ceterus paribus assumption
of the simple two-country trade model.

It is not, of course, unusual to find oneself in the situation of
having to use a less-than-ideal model, especially when working in a rela-

tively unexplored field of research, such as international trade in forest products. Precisely because the model is not ideally suited for the task at hand, I feel that it is essential that all predictions based upon the model be supplemented with an explanation of the likely sources of prediction error, and the probable direction of bias in the predictions. What I am suggesting is that the paper should have provided the reader with additional information explaining the limits of the model, along with a brief discussion of the probable direction and order-of-magnitude of prediction error expected from the use of the comparative-static, two-country trade model to analyze a general reduction in tariffs.

A specific example, drawn from Radcliffe's paper, will perhaps illustrate my point. As part of his estimate of potential additional U.S. imports generated by the reduction in tariffs, Radcliffe predicts important increases in imports of uncoated and coated printing paper and particleboard from Canada. Later, in the section on increases in U.S. exports, he predicts substantial increases in U.S. exports of these same commodities to Canada.

It is not unusual for such two-way flows of the same commodity to take place in international trade. For example, the commodities may not be exact substitutes, or the exports and imports may involve different regions of the trading countries. I suggest, however, that had a more sophisticated model been used, permitting interaction of the major markets, at least a portion of the trade predicted by Radcliffe would have been supplied domestically, either by substitution or by interregional trade within the countries.

With reference to additional U.S. imports generated by the MTN, Radcliffe assumes an infinitely elastic import supply function. This probably is a reasonable simplification in the conventional two-country case, at least for those cases where the additional imports are minor relative to total trade. I question whether an infinite import supply elasticity continues to be a reasonable assumption in the case where all major trading countries are simultaneously reducing their tariffs, thereby, generating a significant aggregate increase in demand. Viewed in isolation, one country's increase in import demand might be negligible in terms of world supply; however, one cannot separate one country's action from the total, because the MTN is contingent upon all countries acting together. Thus, the supply response must be viewed in terms of a simultaneous general expansion in demand for imports by all trading countries. In this situation, the import supply curve faced by the importing country is likely to both shift upward and to be less perfectly elastic. This statement is based upon my belief that at least some of the major exporting countries (that is, the Nordic countries) currently are producing at close to maximum capacity and even a slight increase in output would entail significantly higher marginal costs. If this is the case, the additional U.S. imports generated by the MTN would be less than the level estimated by Radcliffe.

With regard to additional U.S. exports generated by the MTN, I suspect that Radcliffe's estimates understate the potential increases in exports. Much of the previous argument applies here as well; but, in this case, I believe that the controlling factor is the differentials in export capabilities of the major trading countries. As pointed out earlier, Radcliffe's procedure for allocating additional imports among the supplying

ountries assumes that the extra supply will be forthcoming at a constant
ost in all countries. I submit, to the contrary, that extra supply will
e forthcoming in the Nordic countries--and, perhaps, Canada--only at sub-
tantially increasing costs, in contrast to the United States, where the
onstant cost assumption is more reasonable. If this is so, then the
nited States is in a better competitive position to respond to the addi-
ional demand for imports generated by the MTN, and Radcliffe's estimates
nderstate the potential increase in U.S. exports.

For certain forest products, Radcliffe's model probably provides
easonably accurate estimates of the potential changes in trade flows
imulated by the MTN. These would include products dominated by a two-
ountry trade pattern. Examples are hardwood plywood and veneer. In the
ase of more widely traded forest products, such as paper and paperboard
oducts, adequate estimates of the effect of general tariff reductions
an be generated only by recourse to models capable of handling more com-
lex interactions among many countries; namely, the spatial price equili-
rium models. The reactive programming algorithm is particularly well-
uited to handle changes in tariffs.

My bottom line is that I essentially agree with Radcliffe's estimates
 the direction and order-of-magnitude of the trade effects of the MTN.
 suspect, however, that for paper and paperboard products at least, his
stimates overstate the probable increase in U.S. imports and understate
he probable increase in U.S. exports.

Sam Radcliffe is to be congratulated on a fine job, given the very
eal constraints imposed upon him by the model he used, a choice over which
 had very little freedom in view of the absence of a readily available

alternative. My comments certainly should not be interpreted as criticism of Radcliffe's effort, but rather as a commentary on the state-of-the-art in modeling international trade in forest products.

Discussion by Louis Vargha

While I am in general agreement with the conclusions reached by Sam Radcliffe in his paper, I do not agree with some of the specific product conclusions which were reached, and I have some quibbling types of corrections or comments on the paper. The main part of my discussion will deal with the more important of these comments, and the details will be relegated to an appendix to this comment.

In general I think that Radcliffe's conclusions are correct. The MTN did not produce major enough changes in tariff structure to lead to significant changes in patterns or the structure of international trade in forest products. This does not mean that there will not be significant changes, but that they will not be the result of tariff concessions by the United States or its major trading partners. The reductions were, in general, small, and in the case of imports by the United States the bound levels prior to the Tokyo Round provided, with rare exception, no real protection to U.S. industry. One can question the exact degree of change in values which were calculated using 1977 as a base year (which was not a representative year for some odd reason) but overall the impacts, other things being equal, will be small.

The patterns of reductions, and the approaches of the major participants in the negotiations, however, provide some interesting clues concerning what we may expect in the future from our trading partners.

First, the EEC gave us very clear signals of protectionism for its
solid wood and paper industry (there really is no forest products industry
in the EEC). The EEC gave very sparingly indeed, and at that very grudg-
ingly. For the EEC as a whole, and for each one of its member countries,
forest products represent the second largest import bill after oil. As a
result, the community has a balance-of-payments incentive for attempting to
reduce forest products imports, and domestic employment motives for doing
this via import substitution (that is, by domestic production).

To effect import substitution, numerous proposals are being made and
studies conducted, both at the request of the commission of the EEC and by
individual member countries, concerning ways to increase domestic resource
availability, to rationalize collection and distribution methods, and to
rationalize manufacturing. The range of proposals covers secondary fiber
recycling, increasing harvest levels, setting up more effective wood mer-
chandising centers (raw materials collection, sorting and distribution),
and reorganization of paper and paperboard production.

The upshot of all this will be increased government involvement in the
EEC in forest products, and potential direct and indirect subsidies, the
impacts of which probably will not be reduced by the new subsidies code.
The EEC will prove to be a tough market in which to expand product sales.

Japan, on the other hand, made much more significant concessions in
paper and paperboard even though with reluctance and with delayed implementation. Few concessions were made in solid wood products. With these

concessions some fairly clear signals were given to the Japanese pulp and paper industry. The signals were basically to prepare to rationalize and to compete with imports. We'll give you some breathing room; for example, the start of reductions in tariffs on kraft linerboard is timed to coincide with the expiration of an agreement freezing new capacity in Japan and requiring mothballing, conversion, and scrapping of some existing capacity, but over time be ready to change. Clearly, the paper industry was being told that they were not a priority strategic industry for Japan.

On the solid wood side, the message was not there. The industry is more fragmented and--at least in sawmills--more labor-intensive, with a portion of the industry not within the modern sector of Japanese industry. Here, however, there may still be opportunities for further relaxation, as the price and availability of tropical hardwoods could increase constraints on plywood production in Japan and lead to administrative action on tariffs imposed on veneer and plywood. It should be noted in this regard that Japan has been, and may well be in the future, responsive to bilateral pressure, and may administratively reduce applied tariffs even though bound tariffs are not changed.

Canada also gave its pulp and paper industry some messages on being opened up to competitive pressure, and the need to actively rationalize and update eastern Canadian capacity. Surprisingly, the Canadian government made significant concessions in the white paper area (mainly groundwood papers, but also in uncoated free sheet) rather than in kraft linerboard or bleached board products for which U.S. industry had expected more progress. But even in the latter products the case is not closed, but is open to discussion in 1983 during joint talks on progress in implementing concessions. Thus, the atmosphere for liberalized trade in paper and paperboard with Canada is and will be much better than before.

As to conclusions regarding what may be specific market and product impacts, I differ in some important cases with the conclusions in Radcliffe's paper. These differences are caused in large part to some different assumptions. Radcliffe's general conclusions and, I would gather, many of the specific conclusions, should be prefaced with "other things being equal." That preface is appropriate for the general conclusions, but it is not appropriate for the specific conclusions. The following are my major disagreements, and my reasons for them.

Hardwood plywood imports from Korea, Taiwan, and Japan are not likely to increase. They are likely to decrease, and the implications of an increase in total hardwood plywood and veneer consumption in the United States is equally untrue.

The reason is that "other things are not equal" and the supply curve is shifting upward and to the left because of increased raw material and transportation costs to a greater extent than it will shift in the opposite direction because of future tariff reductions. Thus, these are becoming more costly products and, in the face of unchanged demand, total consumption of them would decrease. Other substrates would be substituted for the "blanks" which are the predominant product involved. But in addition to this, there is also a demand shift, with a trend toward reduced use of such overlaid or printed panels in conventional and mobile homes. Thus, it would appear likely that both total consumption and total imports are much more likely to decrease than to increase.

With regard to the specific exporting countries mentioned, it would
ppear unlikely, given their growing domestic requirements for product and
ae desires of the SEALPA countries for more manufacturing, that the ability
f Korea, Taiwan, and Japan to export will increase. Indeed, the reverse is
ore likely.

Mention was made of increases in imports of coated (groundwood) print-
ag paper from Finland and West Germany given the elimination by the United
ates of tariffs on these grades. Although there has been a temporary in-
ease in imports of these grades because of the exceptionally strong demand
r publication-grade papers and lack of domestic U.S. capacity, imports
ould in the future decline to a trickle. Currently, there are several
pansions in these grades in the United States, and domestic capacity will
sufficient to meet domestic demand. Costs in Finland and West Germany
e higher than in the United States and, coupled with increasing transpor-
tion costs, imports from these countries could not compete in the U.S.
rket on the basis of delivered costs.

Similarly, it is not likely that imports from Canada of printing papers
oated or uncoated, groundwood or wood-free) will increase substantially.
e decrease in tariffs is much smaller than price effects in the U.S. mar-
t of a weak Canadian dollar, and declines in the Canadian dollar did not
ad to significant increases in imports for grades which are not in short
pply in the United States. In fact, many of the groundwood mills in
stern Canada that logistically are best positioned to serve the eastern
d midwestern U.S. printing markets are at a substantial cost disadvantage
compared with U.S. mills because of their age and obsolescence and their
chnological dependence on roundwood, which has added to wood costs be-

cause of increasing length of haul. Substantial capital will be required
to reestablish and modernize existing capacity in eastern Canada; expansic
capital will be at a premium and reestablishing of existing eastern Canadian capacity would reduce cost disadvantages but not eliminate them.

As a result, it is likely that U.S. dependence on printing papers (in
cluding newsprint) from Canada will continue to decline, and that the Cana
dian tariff reductions, which proportionately are larger, will lead to increases in exports of printing and writing papers from the United States.

Nor is it likely that U.S. imports of kraft paper and paperboard fror
Canada will increase. The reasons are similar to the ones for printing ar
writing papers.

One could go on, but the point to be made is that specific conclusior
regarding trade potential on even a fairly broad product group basis with
reference to specific trading partners can not be made on an "other things
being equal" basis. This is particularly true in a period in which many
of the basic factors affecting international trade in forest products are
substantially different than they were after World War II when the current
trade in forest products evolved.

Exchange rates are no longer fixed. Energy prices have exploded,
changing the relative importance of the costs of factors of production and
creating severe balance-of-payments problems for many trading countries.
Transportation costs are rising and changing the relationship between fixe
and variable costs of operation, and these changes will impact long-and
short-haul costs. The impact of transportation cost on long-distance
travel in low-value bulk products could be substantial. Wood costs are
rising, and in pulp and paper production many regions (for example, the

Nordic countries) are up against the wall on the use of sawmill residues. New large, high-quality timber reserves such as those in Indonesia are not waiting to be opened. Economies of scale in pulp and papermaking with current technology have been exhausted, and capital costs per ton of new capacity are rising in real terms.

These factors and others (for the depressing list is not exhaustive) are the determinants of the international terms of competition in the future and they are really the significant issues in U.S. forest products trade.

Appendix

A. Specific Points

1. Table 7: not included, but potentially important are:

	Pre-Tokyo	Concessions
19750-2 Printing paper, coated or uncoated in rolls, or rectangular sheets, not less than 22x17 inches, weighing over 18 lb/ream of 432,000 square inches.	12.5%	6.5%
19750-1 Printing paper, coated or uncoated over 18 lb/ream of 432,000 square inches.	12.5%	8.0%

These categories include wood-free printing papers. The second category would include cut-size papers, and the first rolls and folio sheets.

2. Table 8: The pre-Tokyo tariff rate of 5 percent for 47.01-121 and 47.01-126 are "bound" rates; they have not been applied since 1974. The concession to 2.2 percent therefore could not have any effect as concluded in this paper. The Japanese do not plan to apply a tariff, but have kept their options open and a small trading chip in hand. The other tariffs shown for paper and paperboard are applied tariffs (the bound tariffs are higher). It is also important to note that the reductions on 48.01-310, 48.01-321-329, and 48.01-401-420 will not begin until January 1, 1984.

3. Table 9: Full concession rates for the EEC will, at the insistence of the French, be staged in a deux tranches approach. Two rounds (slices) of tariff reductions will be made. Half the concession is to be made January 1, 1983 and 1984. The second half is to be held in reserve

pending review of economic conditions and to be implemented January 1, 1983 and 1987. The EEC negotiators indicated that there would be no situation that they could foresee that would cause the second round of cuts to be canceled (nous verrons).

Newsprint was not shown but a concession was made to reduce the tariff from 7 percent to 4.9 percent and to continue the 1.5 million metric ton duty-free quota.

B. General Quibbles

1. 1977 as a base year for comparative statistics is, in my opinion, a singularly dynamic and unrepresentative base year (there may in fact be no single good base year, a three-year or five-year average may be better). But in 1977 the United States was attempting to reflate while our trading partners in Japan and the EEC were not, thus differentials would exist between the propensity to import in 1977 versus some average propensity to import. In addition, July 1, 1977 saw the elimination of the Swedish subsidy program for maintenance of inventories. This, coupled with an earlier devaluation of the Swedish krona (followed by the Finnish markka) and a subsequent devaluation of the krona in August, caused ripples of market-share distortion in the EEC in 1977 which continued into 1978.

At the same time in one commodity (kraft linerboard) Canadian market shares in the EEC rose as the doomed Labrador liner mill, in a last effort, poured linerboard into the EEC. Thus, on a share basis Canadian exports in 1977 were higher than they were in any previous time, and (with Labrador now being converted to newsprint production) the shares were higher than they possibly could be in the future. Similar types of things happened with Canadian exports of softwood plywood, and, to a lesser extent, lumber.

Although most of this dynamic swirling was concentrated in the EEC, ripples even reached Japan, with Nordic products being shipped in increasing volumes. Thus, 1977 was an interesting year, to say the least, but is hardly my prime candidate for a base year.

2. Infinitely elastic supply curves are a reasonable simplifying assumption for general estimates of percentage changes in imports, given the range of forest product tariff cuts involved in the MTN. But with larger cuts, or if individuals or companies are looking at comparative cost effects, this assumption would be generally invalid even if the production capacity were available (and assuming an average cost of production number) As import volumes increase, the cost of distribution into the importing market (special reference now to finished or semifinished forest products) will tend to increase beyond some level of market penetration. In many cases in the real world we may be at that point. Companies that are evaluating market potential in a specific country should not overlook this phenomenon and view only at c.i.f. point of discharge costs as the true cost to the market when evaluating comparative cost advantage.

PART III

THE ECONOMIC EFFECTS OF LOG EXPORT RESTRICTIONS

WELFARE ECONOMICS AND THE LOG EXPORT POLICY ISSUE

A. Clark Wiseman and Roger Sedjo

Introduction

Economics, according to a standard definition, is a social science which deals with society's use of scarce resources to satisfy material wants. This definition is somewhat limited, particularly when it comes to problems related to economic growth controlling inflation and unemployment, and achieving a fair distribution of society's output among individual households. However, it is a particularly apt description of what microeconomics is all about. This broad area of economics, which embraces most of the problems in the economics of forests, is concerned above all else with efficiency—getting the most out of our resources. Microeconomists abhor waste. Waste means that by using resources differently society can have a larger output with no more resource inputs, including labor power, capital, and natural resources. As far as society's output is concerned, an equivalent situation exists if either (1) resources are used inefficiently; or (2) some resources are simply not utilized; or (3) all resources are used efficiently, but some of the output is set aside to be destroyed. This perhaps explains the microeconomist's fetish for avoiding inefficient use of resources.

In view of this, it is somewhat surprising that so little of the
forest policy work is really directed at this central and basic problem
of economics. A case in point, taken from an area in which we have recently become involved, is work that has been done on U.S. log export restrictions. Economic discussions and analyses of the effects of actual or for-
seeable export restrictions are concerned with effects on employment in domestic work processing relative to log production, effects on the nation's
balance of payments or exchange rate, or both, and of course, effects on
log and lumber market prices and quantities. These are all legitimate, interesting areas for analysis. Notably absent, however, are questions relating to the efficiency of the economy--the gains or losses in the value of
total output that would result from trade controls on logs and the resulting reallocation of society's resources between alternative uses.

The failure to employ welfare analysis more fully in forest policy
analysis is probably due to a large number of factors. However, one of
the more obvious reasons results from the fact that forest economics is a
specific applied area within the broader compass of economics in general.
Most economists in the field have their feet firmly on the ground in the
sense of being concerned with specific problems within the forest industries. Many of these problems relate to optimum resource management and
behavior of prices and quantities for the various forest products. It is
also often the case that economists in applied areas have a distinct
clientele which is interested in specific answers to questions primarily
related to problems lying entirely within the scope of the industry.
Welfare economics is more global in nature and more oriented toward the
use of economics as a tool to aid social decision making rather than as

a tool of management science to enable optimization within firm or industry. It is concerned with society's optimum use of resources and as such, its scope extends beyond the firm or industry to society as a whole. The economist dealing with welfare economics wears the hat of the social scientist rather than that of the market analyst or managerial economist. It is the latter areas that have historically been the primary concern of forest economists, notwithstanding the fact that the analytic problems facing both groups are basically the same. The next section is a brief sketch of the basic postulates and tools of applied welfare analysis. Following this, we give an example of the approach applied to a policy of complete log export prohibition. Finally, we examine a number of implications that arise out of our analysis that relate to welfare, distributional and other outstanding log export policy concerns.

The Underpinnings of Applied Welfare Analysis

The power of applied welfare analysis derives from two important characteristics of demand and supply curves, respectively. On the demand side, the point of departure of the analysis is the postulate that the ordinate of the demand curve shows, for any given quantity, the marginal social value of the good demanded.[1] Since the height of the demand curve shows that the maximum price at which the good can be sold, it is the value

---

[1]The measure of value and cost employed in this paper were developed by Alfred Marshall (1920). Marshall's measure of value is empirically the easiest to deal with, but requires certain restrictive assumptions about consumers' preferences. Other measures have been derived theoretically which do not require these restrictions, but (1) they are difficult to handle empirically; and (2) Willig has shown that under normal conditions the Marshallian measure is a good approximation.

that the buyer(s) place, as measured by the amount they are willing to pay, on an additional unit of the good. In the absence of externalities, the value that buyers place on the marginal unit is also the value to society of that unit.[2] Hence, the height of the demand curve shows the marginal social value of the good. This important feature of the demand curve enables the economist to generate important general inferences regarding changes in social benefits resulting from market price-quantity changes.

Since for any quantity the ordinate of the demand curve shows marginal social value, it follows that the entire area under the demand curve out to a given quantity shows the total value to society of the given quantity. That is, the value of a given amount of a good can be measured by the sum of the maximum amounts the buyers would be willing to pay for each unit taken individually. When we add up the ordinates of the demand curve from the quantity zero out to any specified quantity, we obtain the area between the curve and the horizontal axis.

Buyers generally do not get all of this value, of course, since they pay a price. Suppose that purchasers take a certain quantity at a price shown by the demand curve, that is, they are at a point on the demand curve. Then the amount they pay is simply price times quantity, or the area of a rectangle constructed on the demand curve. Subtracting the amount paid from the total value of the good to buyers yields the area below the demand curve and above the market price. This area was termed by Alfred Marshall, and is still called today, "consumers' surplus." Some writers have moved

---

[2] When externalities are known to exist, it is seldom possible to measure them with precision so that in practice approximate adjustments are made. The only relevant externalities are those of a nonpecuniary type which do not affect market price of the good being analyzed.

ward calling it "buyers' surplus," however, for a reason which points

t an important feature of the surplus. The buyers may in fact not be

e final consumers at all, but instead be producers who use the good as

input in the production of still another good or goods which, in turn,

y or may not be inputs to still other goods. An important finding of

onomic theory is that the full value to all buyers of the good, whether

ey be intermediate or final purchasers, is given by the consumers' sur-

us measure descrived above. It is not correct to add addditional surplu-

s at later processing stages. This has been demonstrated recently by

secarver (1964). The social value of wheat is given by the demand curve

r wheat, and we need not worry about additions or subtractions to this

lue that would be entailed by analyzing the demands for flour or bread.

Many analyses involve alternative prices or price changes, rather than

evaluation of the surplus generated by having the good available rather

an not available. In the case of price changes, there is an increase

decrease in the area between the demand curve and market price. This

ange in consumers' surplus will appear as a trapezoidal area which, for

rposes of comparing different prices, is best thought of as being measur-

horizontally in the sense that it is the area to the left of the demand

rve between the two alternative prices.

As noted above, not all the gain to society from having a given quan-

ty of a good goes to buyers, at least not if they pay for it. A part

the gain goes to the sellers of the good. The gains to the sellers,

wever, is not the revenue shown by the rectangle referred to earlier

at is constructed on the demand curve for a given price-quantity. That

ea shows gross receipts of sellers, and from that their costs must be de-

cted.

One way to approach the matter of costs is to consider a single seller of a good; in this case, a worker who sells labor time in a particular industry. The worker has a supply curve of labor time which reflect his preferences in the face of other opportunities, such as employment elsewhere or leisure time. The height of the worker's supply curve at a given number of hours worked can be regarded as the marginal opportunity cost to the worker. That is, it is the minimum price or wage rate that will just induce the worker to supply one more hour or work. The worker must evaluate the opportunity cost at that hour or work--the value of the foregone opportunity elsewhere--at the minimum amount required to entice him to sacrifice those opportunities. This amount is the wage rate, as shown by his supply curve. It has been shown that for any properly defined supply curve,[3] whether it be the individual's supply of labor or the supply of automobiles, the ordinate of the supply curve shows the marginal social cost of the good produced or sold. Where the good in question is produced by a larger number of different resources owned by different individuals, the ordinate of the supply curve still shows the cost to societ of an additional unit of the good, albeit the costs are in the form of opp ortunities foregone by a large number of owners of different types of resources.

Analogous to the demand curve, the total social cost of a given amour of a good is the total area under the supply curve between quantity zero and the given quantity.[4] Sellers of the good actually receive an amount

---

[3]Following a long controversy in the literature, the basic demonstration of this was provided by Ellis and Fellner (1943).

[4]Again, equality of social cost and private costs exists in the absence of nonpecuniary externalities.

greater than this, namely, the area of a rectangle constructed on the supply curve at a given point (that is, price times quantity). The actual gain to all resources employed in the production and sale of the good is this rectangular area minus the total cost to sellers (the area under the curve or the social cost), leaving as their net gain the area above the supply curve and below market price. This area, known as "producers' surplus" is the net gain to all resource owners--laborers of different kinds and owners of capital and natural resources that are used to produce the good.[5]

A change in the price of a good will result in a change in this surplus, a price rise increasing it and vice-versa. For analyses involving price changes, the change in producers' surplus is measured by a trapezoidal area lying to the left of the supply curve between the two prices in question.

Let us now apply the concepts of producer and consumer surplus to two hypothetical problems. First, consider figure 1. What is the social gain from having the market quantity X_E of the good X sold at the price P_E, as compared with not having the good supplied at all? One way of viewing the problem is to consider the $X_i{}^{th}$ unit. The marginal social value of the unit is $X_i d$, and its marginal social cost is $X_i c$, the gain from producing it is therefore cd. This reasoning applies to all units up to the $X_E{}^{th}$. If we add up all such distances as cd for each unit up to X_E, we

---

[5]Analogous to the demand side of the market, it is incorrect to extend the analysis to the markets for inputs, or of inputs to the inputs. All "prior" surpluses are incorporated in the surplus measured for the market under analysis. The analysis of the division of this surplus among input suppliers is not taken up in this paper.

get the area aEb as the total societal gain. An alternative is to simply
add together the consumers' surplus aEP_E and the producers' surplus bEP_E.
This has the advantage of showing how the gain from having the market
quantity of X produced and sold is split between resource owners and
buyers. This example is important because it illustrates why welfare
analysis is important—indeed, indispensable—as the core of meaningful
cost-benefit or project analysis. Public projects such as transportation,
irrigation, power generation and transmission all involve the basic ques--
tion of societal gains versus costs. Although measurement and other problems abound, the question of which projects are desirable is basically
an application of welfare analysis.

A sort of "cost-benefit analysis" of government policies, e.g., legal
changes, can also be carried out using the welfare approach. As our second
hypothetical example, suppose the government prohibits the export of good
X. In figure 2, the supply and demand curves for X within the country are
shown. If the good can be traded at the world price P_w, that price will
prevail internally. The quantity $P_w A$ is purchased domestically, the quantity $P_w B$ is produced, and the difference, AB, is exported. Now if world
trade is prohibited, the equilibrium will be at Point E, with the domestic
price at P_E. As compared with the trade situation, resources employed in
the production of X lose a surplus equal to area $P_w BEP_E$. Domestic buyers
gain from the lower price, their gain in surplus being $P_w AEP_E$. It will
be noticed that in this example the gain to the one group is less than the
loss to the other. The difference, called a "net welfare loss" or "dead-
weight loss," equal to area ABE, represents economic waste resulting from
the misallocation of resources under the government policy.

Figure 1.

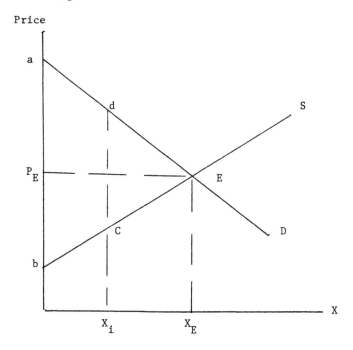

Figure 2.

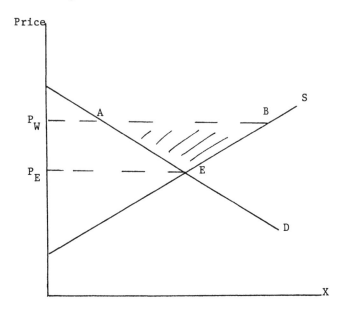

In the next section we show how a model like the one in the last example can be extended to provide estimates of the welfare effects of a complete log export prohibition.

Welfare Analysis of Log Export Restrictions

The welfare approach may be applied to estimate the effects of a hypothetical prohibition of all log exports from the western United States. This was carried out by the authors (Wisemand and Sedjo, 1980) as a means not only of assessing the effects of such a prohibition from the standpoint of overall efficiency (what it does to the size of the economic pie), but also to determine the gains or losses to various affected groups (how the change in the size of the pie is divided up). The analysis is basically an extension of the model given at the end of the last section. Two major complications necessitated the modification of that model into a somewhat more complex one: first, the international trade setting means that some affected groups are not domestic residents. Since the primary concern is with evaluation of welfare effects within the domestic economy, account has to be taken of any gains or losses that are in effect exported to the rest of the world as a result of the policy in question. The second complication results from the fact that logs are themselves an input into the production of more finished products, primarily lumber. As noted earlier, this in itself poses no particular problem for welfare analysis. The problem arises from the fact that lumber is also a traded good. Hence, repercussions in the world lumber market may involve price and quantity changes which substantially affect the analysis. The significance of this effect is evidenced in the discussions and speculations over the extent to which log export restrictions will induce foreign buyers to substitute

mports of lumber for logs; thus reducing or even preventing any resulting
ecline in the price of lumber.

Our analysis involves a synthesis of elements of welfare economics,
nternational trade, and the theory of derived demand.[6] The derived
emand concept is the vehicle through which log and lumber markets are
elated in our analysis, and it will be necessary to explicate the derived
emand model before proceeding.

In figure 3, the long-run supply curves of logs and log processing
re depicted and labeled S_1 and S_p respectively. The latter can be thought
f as a composite of the processes that convert logs to lumber. In the
nternational trade literature this composite is referred to as the "value
dded industry." As shown, they both slope upward, indicating the analysis
olds equally well for one or both industries being constant or decreasing
ost industries. The supply curve of lumber S_L is the vertical summation
f the log and processing supply curves: it is the locus of prices which
ll just cover both components of the cost of lumber production. Notice
hat the vertical distance between S_1 and S_L represents unit processing
osts.[7]

Again, figure 4 shows the supply of processing S_p. This is subtracted
ertically from the demand curve for lumber D_L to obtain the "derived

---

[6]The derived demand model, like the concept of surplus discussed above
s a product of the genuis of Alfred Marshall (1920, book V, chapter XI).
r obvious empirical reasons, the fixed factor proportions version of this
odel is employed in the present paper.

[7]The standard unit of lumber having been decided upon, units of logs
ay be defined as the quantity required per unit of lumber. Similarly, a
nit of processing" is the amount per unit of lumber; hence, prices and
uantities can be depicted as shown in the figures.

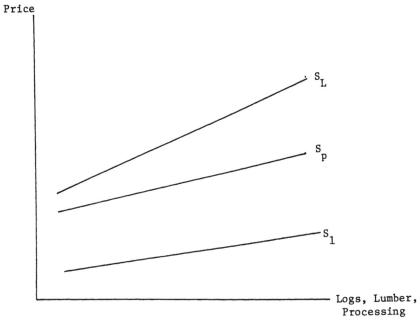

Figure 3.

Price

S_L

S_p

S_1

Logs, Lumber,
Processing

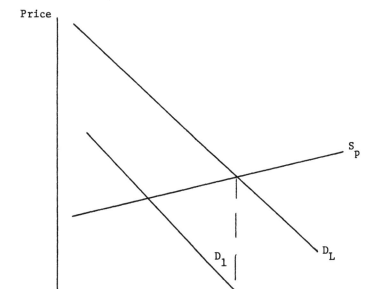

Figure 4.

Price

S_p

D_L

D_1

Logs, Lumber,
Processing

demand" for logs D_1. The derived demand concept is necessary because
there is no significant want-satisfying quality in logs themselves that
makes them an object of man's desires. Instead, logs are demanded by
processers who are willing to pay, for a given quantity, no more than a
price which is equal to the "gap" between lumber price and the unit cost
of processing.

Combining parts of figures 3 and 4 allows us in figure 5 to represent
the log and lumber markets simultaneously. The equilibria in the two
markets are at the two points of intersection of the demand and supply
curves. Now let us suppose that both logs and lumber are exported by this
country. In figure 6 the world price of logs, u, is above the autarky
equilibrium price in the domestic economy, resulting in exports of logs
equal to yz. The uy logs produced are processed at cost yx, so that the
supply price of the quantity vx of lumber produces is v.[8] Our treatment
of the trade situation in lumber is slightly different from that for logs.
Notice, in particular, that we have surreptitiously added a small line,
D_L^F, in the diagram. That line represents foreign demand for domestic
lumber, and we must now reveal that our demand curve for lumber D_L is in
fact a _total_ demand curve which includes both domestic and foreign demand.
D_L is a horizontal summation of foreign and domestic demands. The latter,
although not shown explicitly, is the horizontal distance between D_L^F and
D_L. As shown, at the price v, vw lumber is exported and wx is sold domestically.

---

[8]The supply curve of lumber is drawn assuming no log trade, giving
the appearance that point X is not on a supply curve. In fact, S_L shifts
to the left through point X when log exports are yz.

Figure 5.

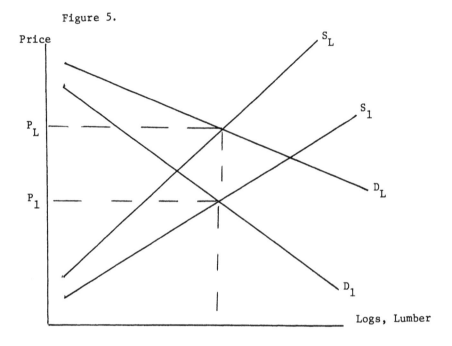

Figure 6.

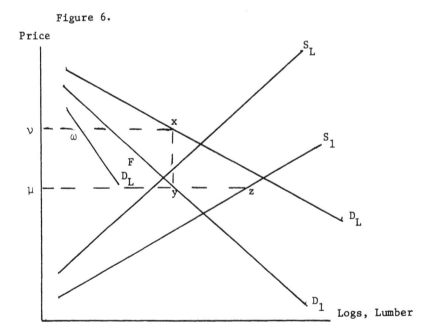

Well-equipped with our analytical tool kit, we may now examine the theoretical effects of a log export prohibition. This is done in our final diagram (figure 7). The prohibition of log exports results in equilibria in log and lumber markets where the respective supply and demand curves intersect. Prices fall from u and v to u' and v'. In the log market there is a loss in producers' surplus and a lesser gain in buyers' surplus, leaving the net social welfare loss equal to the shaded triangular area. The gain to buyers goes both to owners of resources employed in the processing industry and to buyers of lumber. The part of this gain accruing to buyers of lumber is shown by the area of a trapezoid constructed on D_L between v and v'. But, observe that a part of this gain is in fact a gain to nondomestic lumber buyers whose consumers' surplus rises by the shaded area shown toward the top of the diagram. This gain is a reflection of the fact that foreign buyers benefit from purchasing more lumber at a lower price. The net social welfare loss due to the policy is the sum of the two cross-hatched areas. Hence, in an international setting, the traditional view that the net change in welfare can be shown entirely in the input market must give way to the possibility that some of the surplus change in effect may be exported to the rest of the world. This is why it was necessary in our discussion to extend our focus beyond the log market and to relate it to the lumber market via the derived demand model.

As shown, the analysis assumes that the foreign demand for lumber is unaffected by the discontinuance of log exports. There is reason to expect that foreclosure of a log supply source will tend to shift foreign demand for domestic lumber to the right. A rightward shift in D_L^F will shift both D_L and D_1 to the right, so the log and lumber prices will not

Figure 7.

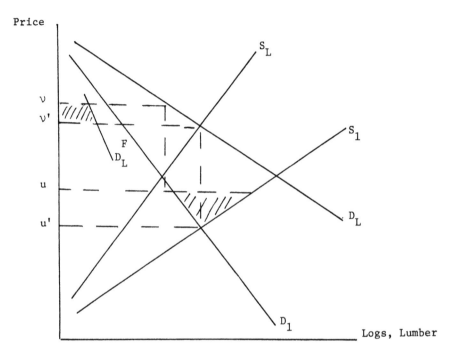

all to the extent depicted. Using the model above, we have shown elsewhere that it is possible to bracket a range of values within which price,
quantity, and surplus changes must fall (Wiseman and Sedjo, 1980).

The relevant effects of a log export prohibition have been estimated
using market data on prices and quantities, technical data on the value
added of the domestic processing industry, econometric estimates of
demand and supply elasticities, and certain relations between elasticities
of the various functions imposed by the model itself. The Pacific Coast
region (PCR) is the area selected for analysis.[9] Our estimates are given
for two scenarios: the one depicted in figure 7 where there is no "feedback" of increased nondomestic lumber demand resulting from the policy,
and the other theoretical limit where there is a "maximum feedback" of
increased nondomestic demand. Estimates of the market equilibrium effects
are given in table 1 and the corresponding welfare effects in table 2.

Issues and Implications

There are a number of broad economic and social issues that commonly
arise and should be addressed when discussing the desirability of a log
export restriction and upon which this analysis has implications. These
include:

1. The overall social welfare question

2. The prices of logs and lumber

---

[9] This region, comprising Washington, Oregon, and California, was
selected for analysis because the objective of past and proposed legislation is to restrict log exports from this region to the Far East, primarily
Japan. Politically, pressure both for and against legislation comes largely from legislators and groups within the PCR, making it a logical unit for
analysis from a political economy served by selecting that region, rather
than a larger or smaller geographic entity.

Table 1. Estimated Effects on PCR Log and Lumber Markets Resulting From
Complete Log Export Prohibition

	No feedback of foreign lumber demand in PCR lumber market		Maximum possible feedba‹ of foreign lumber demaɴ in PCR lumber market	
	Percentage change	Absolute change[a]	Percentage change	Absolut‹ change‹
Log market				
Price	-12.0		-3.7	
Quantity				
Production	-4.0		-1.2	
Domestic sales	+13.1		+16.3	
Exports	-100.0		-100.0	
Lumber market				
Price	-3.7	-$7.70	+.4	+$.80
Quantity				
Production	+13.1	+2140	+16.3	+2660
Domestic sales	+1.3	+170	-.2	-20
Exports	+57.5	+1970	+92.4	+2680

Note: Basis of comparison is current regime of tariff and other tra‹
barriers, but with no U.S. log export restrictions. The estimates reflecᴛ
conditions and prices obtained circa 1976.

[a]Price is dollars per 1,000 board-feet. Quantity is millions of
board-feet. Calculations in the text give log price in dollars per log
quantity required per 1,000 lumber board-feet and are not reported here.
Log units, similarly defined, are not reported. Percentage changes for
logs may be used to obtain absolute changes per standard log unit and uniᴛ
price.

3. The employment effects

4. The effect upon the harvest in regional forests

5. The balance-of-payments effects

6. The redistribution effects

Let us briefly discuss the implications of our analysis for these is‹

sues. First, our analysis suggests that the net regional welfare losses

Table 2. Estimated Annual Welfare Effects on PCR Market Participants Resulting from Complete Log Export Prohibition

(millions of dollars)

	No feedback of foreign lumber demand in PCR lumber market	Maximum possible feedback of foreign lumber demand in PCR lumber market
Owners of resources employed in PCR		
Log production	-209.8	-66.2
Processing logs to lumber	+60.3	+76.2
PRC purchasers of lumber	+99.5	-10.9
Net welfare gain to PCR	-50.0	-0.9

Note: Basis of comparison is current regime of tariff and other trade barriers, but with no U.S. log export restrictions. The estimates reflect conditions and prices obtained circa 1976.

associated with a log prohibition range from a maximum of $50 million per annum to a minimum of about $1 million per annum. The extent of the losses depends upon the degree of feedback, that is, increased regional demand by the Japanese for lumber to offset her log resource losses because of the prohibition. To the extent that the feedback is large, the net regional welfare losses will decline.

Second, the analysis predicts an unambiguous decline in log prices as the result of a prohibition. However, a decline in lumber prices is less certain, and if it does occur, we predict that it will be quite modest. It should also be noted that if a price decline is experienced and is relatively large, the implication is that the net welfare loss of the region is also toward the high end of our range of estimates.

Next, although we have not estimated employment directly, one can get an idea of the industry effects from the production effect. Our analysis suggests an unambiguous increase in regional processing and hence processing employment. However, total log production will decline resulting in employment reductions in production and associated activities. Furthermore, the decline in log price can be expected to adversely impact upon forest growing activities and hence create additional adverse employment impacts. Even if the forest industry broadly defined experiences employment increases, increase in industry employment should not be confused with increases in regional economy employment. As Stevens (1979) has shown, many wood workers have an unexpectedly high degree of job mobility and move in and out of the industry rather readily. The overall levels of employment are determined by monetary and fiscal factors. The restriction will affect large the distribution of employment within the region.

Some groups are concerned about the effect of the lack of a prohibition upon the level of cut in regional forests. Our analysis suggests that the cut would be reduced but that the reduction would be relatively modest.

Further, while our analysis is not directed at the balance of payments issues it does provide some indirect information and a clarified perspective in regard to this question. First, exports of logs would be lost and hence the foreign exchange earning would be forgone. However, foreign exchange would be earned either to the extent that the region increased its international exports of lumber or increases in domestic lumber production replaced lumber previously imported from international suppliers.

Finally, our analysis indicates that the redistribution implications of a prohibition are certainly great. The major gains and losses from a log export restriction for the active participants involve primarily the redistributional consequences of the restriction. The analysis predicts that losses will always be experienced by producers of logs, i.e., the land, labor, and capital utilized and specialized in log production. Our estimates of the losses range from $66.2 million to $209.8 million per annum. This is equivalent to a one time loss (using a 10 percent discount rate) of between $662 million to $2.1 billion. Unambiguous gains are predicted for processors ranging from $60.3 million to $76.2 million per annum. The third affected group are the regional lumber purchasers. Our analysis indicates that the impact upon them can be either positive or negative. However, to the extent that lumber prices fall and purchaser gains are large, our analysis shows that this situation would be associated with rather large regional net welfare losses. It is not surprising therefore that producer-related groups, who have a great deal to lose from a prohibition, would be expected to oppose a prohibition while processor groups, who anticipate gains, would be expected to support a prohibition. Finally, consumer groups might, rightly, be confused as to their expected gain or loss from a prohibition.

REFERENCES

Ellis, Howard S. and William Fellner. 1943. "External Economies and Dis-economies," _American Economic Review_, vol. 33, no. 3, pp. 493-55.

Marshall, Alfred. 1920. _Principles of Economics_, 8th edition (New York: Macmillan).

Stevens, Joe B. 1979. "Six Views About a Wood Products Labor Force, Most of Which May be Wrong," _Journal of Forestry_, vol. 77 no. 11 (November).

Willig, Robert D. 1976. "Consumer's Surplus Without Apology," _American Economic Review_, vol. 66, no. 4, pp. 589-597.

_____. 1979. "Consumer's Surplus Without Apology: A Reply," _American Economic Review_, vol. 69, no. 3, pp. 469-74.

Wisecarver, Daniel. 1974. "The Social Costs of Input-Market Distortions," _American Economic Review_, vol. 64, no. 3, pp. 359-72.

Wiseman, A. Clark and Roger A. Sedjo. 1980. "Log Export Prohibition: Efficiency and Distribution Implications." Unpublished manuscript.

Discussion by Philip Cartwright

The use of welfare economics to analyze the effects of a complete log
export restriction, or even a partial export restriction which reduces the
volume of log exports from that which would prevail in a "free" market, is
useful analytical tool for examining the economic benefit or loss to be
derived from such a foreign trade policy. As Sedjo and Wiseman have indicated in their paper, the welfare approach is designed to indicate
whether resources are being allocated among alternative uses in a manner
which maximizes the value of output from the resources. The application
by Sedjo and Wiseman of this analytical tool to the production of logs and
lumber for domestic consumption and export in the face of an embargo on all
logs provides a sort of first approximation to the possible losses and/or
gains in welfare which might result for the United States and the redistribution of welfare from the log production industry to the log processing
industry.

Welfare economics is based on Marshallian economics which assumes
perfectly competitive markets for both factor inputs and product outputs.
In a closed economy if all factors and all goods are priced competitively
then prices and outputs will be determined by the intersections of supply
and demand functions and resources will be allocated efficiently among the
production of alternative products. The composition of output will

reflect consumers' choices as well as maximize the returns to the owners of resources, including labor, as a group. Any restraint of competition in product or factor markets through monopolistic elements or governmenta restrictions will alter the allocation of resources and reduce the total social welfare. Sedjo and Wiseman implicitly assume that the only restra of competitive markets for logs and lumber, and, for that matter for all other outputs is imposition of a complete embargo on the export of logs. Both logs and lumber are sold in perfectly competitive markets where pric is determined by the equilibrium of the supply function with the demand function, and lumber and logs are homogeneous in all markets. Without delineation of products and no dominant sellers or buyers such competitio might exist. However, in the log market the federal government is certai ly a dominant seller, able if it so chose to affect the supply and supply price. In the case of a partial log embargo which shifts the supply function, the embargo may, as Dowdle suggests in his paper, have the effe of creating a dominant private seller who gains monopoly power in the export market to his advantage but alters the allocation of resources both in the United States and in importing countries toward less efficiency.

On the buyer's side it may be that the U.S. output of lumber may not be completely homogeneous with lumber desired by foreign consumers which has the effect of creating more than one commodity market for lumber and seriously affects the model and results to which the analysis of Sedjo an Wiseman lead.

Economic efficiency in the allocation of resources to alternative outputs in a closed economy changes significantly when the system is opened and trade with other economies is considered. The resources base and consumer markets are expanded to include all trading partners.

Efficiency in the allocation of resources in each economy is now dictated

by another principle espoused by Alfred Marshall and earlier writers,

namely the principle of "comparative advantage." Efficiency in the allo-

cation of resources in the economies of each of the trading partners,

requires that resources be allocated in each economy toward the production

of those products in which each country has a comparative advantage.

Products for which a country has a comparative disadvantage are obtained

through trade and an equilibrium in the amount of products traded and

the balance of trade (actually the balance of payments) between the trading

partners is realized through the exchange rate of the currencies of the

countries. Any restrictions to trade such as tariffs, embargoes, and so

forth will result in misallocation of resources in each trading partner's

economy, reducing the total return to resources in both countries—in ef-

fect, reducing total welfare.

Applying the principle of comparative advantage to the log and lumber

industries in the United States, it seems obvious that the United States

has a comparative advantage in the production of logs relative to Far East-

ern countries who are willing to buy over 2.5 billion board-feet at prices

above the U.S. domestic price. It is not clear, however, that the United

States has any particular comparative advantage in the processing of logs

into lumber. Our trading partners in the Far East clearly find that pro-

cessing logs into lumber in their own economies yields higher returns than

purchasing logs processed into lumber by U.S. firms. The reason for this

higher return may be due to a difference between the lumber products desi-

red in the foreign country and lumber products produced by U.S. processing

firms. If this is the case, the assumption by Sedjo and Wiseman of homogeneity in lumber products and the assumed foreign demand function for U.S.
lumber leads to erroneous conclusions in both their limiting cases. Another possible reason for foreign countries finding the processing of logs
into lumber advantageous may be in the degree of labor intensity of this
industry and the ratio of labor to other resources in their countries as
compared with this ratio in the United States. A country with a relatively
high ratio of labor to other resources will find that its comparative advantage in trade will be in those products whose production is labor intensive, while the reverse is true for countries with a relatively low ratio
of labor to other resources, such as the United States. In the model
Sedjo and Wiseman present, resources devoted to lumber production in the
United States are substantially increased in both limiting cases. These
additional resources are not drawn entirely from the log-producing industry
but from other industries. It is implicitly assumed from their choice of
supply curves that the transferred resources have a higher return in the
production of lumber than in the industries from which they were transferred. Sedjo and Wiseman present only a partial equilibrium analysis both
in terms of domestic production and foreign trade, whereas the total welfare affects of the resource allocations occuring would require an examination of all other industries in the United States and the rest of the
world.

In any event, the embargo of log exports will presumably interfere
with the allocation of resources according to comparative advantage, reducing total returns to resources in the world economy. If this were not
the case, the embargo would not be needed, and the free trade market would

produce the reallocation proposed by proponents of the embargo. Of course, the proponents of the embargo are not interested in the welfare of the world but only that of the United States. Sedjo and Wiseman, in turn, are only concerned with what happens to welfare in the United States. Sedjo and Wiseman conclude that welfare in the United States would decline, as a result of the embargo in all cases. This conclusions rests on the particular elasticities of supply and demand chosen for the world. I am inclined to believe that they are correct, especially in the event of no feedback of foreign lumber demand in the U.S. market.

One can imagine, however, a situation where an embargo of a raw material could improve the terms of trade and welfare for one country against another trading partner. For example, if the OPEC countries embargoed crude oil to the United States and permitted the export of refined oil products only, the United States would have no alternative sources of substitutes for crude oil at competitive prices. We would thus be forced to devote more resources to obtain oil products from the OPEC countries or from alternative domestic substitutes. The terms of trade would move against the United States. Welfare in the combined countries would fall, but the loss would be entirely in the United States. In fact, the loss in the United States would be greater than the total loss as welfare, that is, producer's surplus would improve in the OPEC countries. But this example depends on the particular monopoly power of the OPEC countries over the U.S. market.

It is not clear that the United States has a similar monopoly power over foreign countries who now import logs from the United States. These countries may find alternative substitutes for U.S. logs, in the event of an embargo, which although purchased at a higher supply price may be pro-cessed into lumber at a lower price than the U.S. export price of lumber. Moreover, if logs are not available to the foreign countries alternative sources other than the United States of the particular type lumber desired may be available. Again, examination of these broader questions are nece-ssary before the conclusions of Sedjo and Wiseman can be accepted.

In the analysis of Sedjo and Wiseman, the welfare gains to the proces-sing industry and, in one case, to the purchasers of lumber, result from the diversion of logs from the export market to the domestic market with a consequent fall in price. The result is a consequence of the particular supply function of logs which is assumed. This supply function, it ap-pears to me, is based on the artifically imposed restraint of the dominant supplier of timber to the log market, the Forest Service. It is well known that the Forest Service by its economically inefficient management practice of an even-flow cutting policy in the face of an excessive inventory of timber on public lands is constraining the supply function and raising the price of logs. If there are welfare gains to the domestic processing in-dustry and to buyers of lumber from a diversion of logs from export to the domestic processing industry, the same gains could be achieved by increas-ing the supply of timber from public lands.

The results which Sedjo and Wiseman indicate in the limiting case of "no feedback of foreign lumber demand" show an increase in exports of lumber of 1,970 million board feet, presumably as a result of the 3.7 percent price decrease of lumber. The lack of homogeneity of domestically produced lumber with that desired by foreign buyers together with the in-ability or unprofitability for U.S. processors to produce the lumber to

reign specifications, appear to me to make this result unlikely. The
asticity of the foreign demand for U.S. lumber appears to me to be much
wer than that assumed in the model.

Most of the concerns I have raised with respect to the Sedjo and Wise-
n model if they could be incorporated would, I believe, produce greater
sses in welfare to the United States from a complete embargo on logs.
t only would welfare in the log-producing industry decline, but welfare
 U.S. consumers of all other products imported from the countries to whom
 presently sell logs would decline.

U.S.-JAPANESE LOG TRADE--EFFECT OF A BAN

Richard Haynes, David Darr, and Darius Adams

The decade-long debate over softwood log export policy has at variou
times focused on price and related impacts of restrictions on log exports
Previous studies (Wiener, 1973; Stanford Research Institute, 1974; Clawso
1975; Haynes, 1976; Sedjo and Wiseman, 1979; Darr and co-workers, n.d.)
have, in general, concluded that a ban on log exports would result in (1)
increased purchases of lumber from Canada and the United States by Japan;
(2) only a limited effect on U.S. end-product prices; and (3) the stumpag
market being affected more than the product market.

The basis for concern over the impacts of a ban on log exports has
centered on the price and availability of timber in the Pacific Northwest
and on the price of softwood construction materials in U.S. markets. Ind
viduals and organizations in favor of log exports have argued that higher
stumpage prices caused by the export market are an incentive for more intensive forest management which increases timber supply and leads to bett
timber utilization. They also argue that log exports reduce pressure on
domestic product prices (assuming that U.S. and Canadian producers would
export substantial volumes of lumber if they are unable to export logs).

Those against log exports have argued that higher stumpage prices increase competition for timber in the Pacific Northwest, making producers in the region less competitive in domestic markets as compared with other U.S. and Canadian producers. Those against log exports have also argued that higher stumpage prices lead to higher lumber prices and, ultimately, to higher costs for housing in the United States. Whether product prices actually fall, following a ban, as implied by the opponents of log exports, depends on the responses of U.S. domestic and export markets.

In this paper, the potential effects of a ban on log exports are again examined but in more detail than in past studies. This examination includes impacts on prices and production as well as welfare implications associated with a ban on log exports. Impacts in the stumpage and product (lumber and plywood) markets are discussed at both national and regional levels. The discussion of impacts includes Canada, since Canadian lumber producers have the potential to offset changes in the U.S. domestic market.

We do not attempt to predict the actual sequence of events following a ban on log exports. Rather, we use the Timber Assessment Market Model (Adams and Haynes, in press) to stimulate the impacts (until 1990) associated with two possible sequence of events or scenarios representing different extreme assumptions about the response of Canadian and U.S. lumber exporters following a ban on log exports. Both scenarios assume that Japan replaces logs imported from the United States with the equivalent volume of lumber (4.42 billion board-feet)--assumes log exports from the West Coast average 2.6 billion board-feet per year, and the Japanese recover 1.7 board-feet of lumber per board-foot of log. The first scenario portrays a situation in which Japan buys the entire 4.42 billion board-feet of lumber from U.S. West Coast producers. The second scenario assumes that Japan purchases the entire lumber volume from Canadian producers. Lumber export levels are shown in table 1 for each scenario, as well as for a third scenario that is used as a base for comparisons. This latter scenario assumes that the volume of log exports from the U.S. West Coast remains at 2.6 billion board-feet per year during the decade, 1980-90.

The Market Structure

The Timber Assessment Market Model (TAMM) (Adams and Haynes, in press) is a spatial equilibrium model and, as such, represents the workings of interactive competitive markets. The model consists of sets of supply and demand regulations specified for nine supply regions (including Canada) and six demand regions. The stumpage market is solved simultaneously with the product market to determine the level of timber harvest, by owner group, and the price of stumpage. Stumpage demand is composed of the sum of roundwood inputs requirements for each major product category: lumber, plywood, pulp products, miscellaneous products, fuelwood, and log exports. Stumpage supply is the summation of harvests in public and private ownerships.

A ban on log exports is simulated, using TAMM, by reducing projections of log exports to zero. The primary effect of a ban on log exports is to reduce the demand for available timber which lowers stumpage prices. The greatest reduction in stumpage price occurs in the two log-exporting regions (the Pacific Southwest and Pacific Northwest, the western side of the Cascade Range), although small changes take place in other regions because of interregional dependencies. Lower stumpage prices reduce harvests on

Table 1. Lumber Export Volumes Assumed for the United States and Canada by Scenario, for Selected Years 1976–90

(million board-feet)

Year	Base scenario		Scenario 1		Scenario 2	
	United States	Canada[a]	United States	Canada[a]	United States	Canada[a]
1976[b]	1,481	1,137	1,481	1,137	1,481	1,137
1980	1,369	1,167	4,789	1,167	1,369	6,087
1985	1,485	1,833	5,905	1,833	1,485	6,253
1990	1,600	2,000	6,020	2,000	1,600	6,420

[a]Canadian exports·to countries other than the United States.

[b]Average of 1972–76.

private lands and the drain in timber inventories. They also reduce the incentives for intensive management resulting eventually in lower timber inventories. Lower stumpage prices would also be reflected in production costs and, eventually, in prices for softwood lumber and plywood. Lower costs of production would increase processors' profits, providing an incentive for capacity expansions. In the ban on log exports, lumber and plywood production would expand in the Pacific Southwest and Northwest (west side) regions at the expense of other regions. Coincident with this expansion of lumber and plywood production, the residue-based western paper industry would face declining pressure to substitute roundwood for residues further contributing to available roundwood supplies.

Lower costs of producing U.S. lumber improve the relative competitiveness of U.S. regions, which affects the levels of lumber imports from Canada.

This sequence of events assumes that the processing capacity in the domestic U.S. regions changes in response to cost conditions. In simulating both scenarios, we rely on the capacity adjustment mechanism in TAMM to control the rate of capacity response to changing cost conditions. Essentially, this adjustment mechanism attempts to maintain, in each region, reasonable profit levels, averaging $12.50 per 1,000 board-feet for lumber and $40 per 1,000 square feet, 3/8-inch basis, for plywood. Lower stumpage prices resulting from restrictions on log exports should increase profit levels and serve as an inducement for capacity expansion. Production costs (particularly for stumpage) should rise as capacity expands, leading eventually to lower profits that act as an incentive to reduce capacity.

Price and Production Impacts of the Two Scenarios

esponse Measured by National Aggregates

A ban on log exports, in terms of our representation, amounts to re-
ucing the derived demand for stumpage. The composition of the derived
emand changes are shown in table 2 for the United States.

In both scenarios, log export restrictions are assumed to decline to
ero by 1980, thus reducing the demand for stumpage by 491 million cubic
eet. Nearly all the reduction is in the Pacific Northwest (west side)
egion. Production of lumber would expand in both scenarios, although for
ifferent reasons. In the first scenario, lumber production would expand
s a consequence of increased U.S. lumber exports to Japan, whereas in the
econd scenario, U.S. lumber production would expand to offset a reduction
n lumber imports from Canada. Plywood production would expand slightly
ecause of lower costs in both scenarios.

Increased lumber and plywood production would be accompanied by in-
reased availability of residues, especially in the Pacific Northwest (west
ide) region, reducing the amount of roundwood used in the manufacture of
ulp.

The price and consumption differences between the base scenario and
he two alternative scenarios are shown in table 3. All prices are deflat-
d by the wholesale price index (1967=100). These differences are, per-
aps, smaller than some people would expect given the relatively large size
f the assumed log export volume. However, the 1990 net change in demand
s the determining factor and, as a matter of perspective, amounts to only
.7 and 1.8 percent of total harvest for the two scenarios, respectively.

Table 2. Composition of Net Changes in U.S. Wood Demand Resulting from a
         Ban on Log Exports, by Scenario, 1980 and 1990

(million cubic feet, roundwood equivalent)

| | Changes from base simulation levels | | | |
| | Scenario 1 | | Scenario 2 | |
Change	1980	1990	1980	1990
Source of change				
Reduction in log exports	-491	-491	-491	-491
Change in production of products	343	556	293	516
Shift in pulpwood mix	-83	-260	-77	-229
Net change in U.S. demand	-231	-195	-275	-204

In the base scenario, log exports comprised 4.2 percent of the total 1990

harvest.

Lumber prices increase in the near term but decline to below base

scenario levels after 1985 (table 3). This results from rapidly expanding

processing capacity in the Pacific Northwest (west side) region. Plywood

production also increases in response to lower wood costs, thereby de-

creasing plywood prices.

Regional Responses

The responses to stumpage prices to a ban on log exports are shown in

table 4. The largest impacts would be in the Pacific Northwest region,

although stumpage prices are affected throughout the west. In the Pacific

Northwest (west side) region, the derived demand for timber would be re-

duced roughly 7 percent for the two scenarios. The reduction in west side

Table 3. Consumption and Price Changes from Base Simulation Levels, by
          Scenario, 1980, 1985, 1990

Consumption and price	Scenario 1			Scenario 2		
	1980	1985	1990	1980	1985	1990
U.S. lumber consumption (million board-feet)	-854	1	129	-660	-6	124
U.S. plywood consumption (million square feet)	254	305	286	297	314	327
Lumber wholesale price index (1967=100.0)	9.87	-0.4	-1.5	6.17	-1.3	-3.7
Plywood wholesale price index (1967=100.0)	-3.35	-4.7	-4.3	-4.15	-5.2	-4.6

stumpage prices reflects in part the low responsiveness of private owners.
Price responsiveness measures the percentage change in volume divided by
the percentage change in price and is equal to 0.26 for forest industry and
0.09 for the private owners.

The decrease in stumpage prices accompanied by increases in the demand
for domestically produced lumber would result in substantial changes in regional lumber production (table 5) which are not apparent in table 2.
Lower stumpage prices in the Pacific Northwest (west side) region would
improve the profitability of the region relative to other regions and would
lead, in both scenarios, to an immediate expansion of production. This
expansion would slow and later decline as stumpage prices increased. The
production response in other regions varies--the largest changes are in
Canadian production.

Table 4. Projected Stumpage Prices by Scenario, 1980, 1985 and 1990

(dollars per thousand board-feet, Scribner scale)

Timber supply region	Scenario (1980)			Scenario (1985)			Scenario (1990)		
	Base	1	2	Base	1	2	Base	1	2
North	27	27	27	29	29	28	30	30	30
South	56	57	56	72	72	71	85	85	85
Rocky Mountain	28	36	37	42	41	42	53	52	52
Pacific Southwest	39	38	36	55	51	52	71	68	65
Pacific Northwest									
West side	76	39	36	88	52	49	104	68	63
East side	31	39	36	48	48	52	59	56	58

Table 5. Projected Regional Lumber Production and Imports from Canada, by Scenario, 1980, 1985 and 1990 (billion board-feet)

Timber supply region	Scenario (1980)			Scenario (1985)			Scenario (1990)		
	Base	1	2	Base	1	2	Base	1	2
North	1.2	1.2	1.2	1.2	1.2	1.2	1.3	1.3	1.3
South	8.8	9.2	10.2	10.2	10.2	10.1	10.7	10.6	10.6
Rocky Mountains	5.2	5.3	5.3	5.4	5.3	5.4	5.4	5.5	5.5
Pacific Southwest	5.5	5.6	5.5	5.5	5.7	5.7	5.6	5.7	5.7
Pacific Northwest									
West side	8.2	9.7	9.6	6.7	9.6	9.5	5.5	9.3	9.0
East side	2.8	2.9	2.9	2.9	3.0	3.0	2.9	2.9	3.0
Canada									
Total production	14.2	15.6	16.2	16.9	18.2	18.5	20.1	20.7	21.0
Imports to the United States	7.7	9.0	5.2	10.0	11.3	7.1	12.7	13.3	9.2

Canadian lumber production would vary in the short run (1980) with each scenario, but by 1985, little difference would exist between the two scenarios that assume a ban on log exports. Lumber imports to the United States, however, would be substantially affected by either scenario and would be a key factor in determining the effects of a ban on U.S. markets. In either scenario, imports from Canada would change to offset the change in the U.S. domestic market. In the first scenario where the United States would increase lumber exports, lumber imports from Canada would be increased to partially offset changes in the domestic market. In the second scenario, Canadian lumber imports would decline below the base scenario level and domestic production would increase to offset the decline.

Lower stumpage prices in the Pacific Northwest (west side) region also would result in an expansion of regional plywood production at the expense of other western regions and the South. This expansion, however, would be able only to slow the long-term shift in plywood capacity from the West to the South, and by 1990, the U.S. production share attributable to the South would be reduced by only 1 percent.

Distribution of Impacts

Analysis of only the effects of a ban on log exports on prices and production levels ignores the question of who gains and who losses from the change in policy. Proponents of a ban argue that consumers and processors gain from lower prices, whereas opponents argue that a ban would decrease returns to stumpage producers that would offset gains to consumers and processors. Estimates of losses and gains for 1990 are shown in table 6.

Table 6. Changes Relative to the Base Scenario in Consumer Expenditures, Processor Revenues, and Stumpage Producers' Revenues in 1990 for the Two Scenarios Representing a Ban of Log Exports (millions of dollars)

Region	Scenario 1			Scenario 2		
	Consumers	Processors	Stumpage producers	Consumers	Processors	Stumpage producers
Pacific Northwest						
West side	12.8	322.0	-482.8	11.2	308.7	-537.3
East side		-7.4	-7.6		-11.1	-.1
Pacific Southwest	16.4[a]	-.1	-15.1	5.5[a]	-18.7	-28.1
Rocky Mountain	6.2[b]	-6.5	-3.6	7.4[b]	-11.3	-2.2
South Central	24.6	-22.8	-8.0	34.0	-30.8	-4.8
Southeast		-11.7	-3.4		-17.7	-4.4
North	45.2	c	c	53.0	c	c
Total	105.2	273.6	-520.5	111.0	219.0	-577.9

[a]Includes Nevada, Arizona, New Mexico, and California.

[b]Excludes Nevada, Arizona and New Mexico.

[c]Not available.

The gain to consumers was measured by the overall reduction in consumer outlays. This was estimated as the product of the initial level of lumber and plywood consumption multiplied by the price reduction brought about by a ban on log exports. This method followed McKillop's (1974) suggestion that the change in consumer expenditures be adjusted for the potential reduction in outlays for substitute products. The increase in lumber and plywood output evaluated at the new price is assumed to be an appropriate measure of the reduction in outlays for substitute products. Gains to processors were computed as the difference in total profits between each scenario representing a ban and the base scenario. The losses to stumpage producers were computed as the difference in stumpage price multiplied by the average stumpage quantity (between the base and the scenario representing a ban).

Processors gain more than consumers, and nearly the entire gain would be to lumber processors in the Pacific Northwest (west side) region. Some of the west side processor gains would be offset by declines in processor revenues in other regions. Consumers would gain somewhat less than processors. Slightly more than half the consumer gains would be attributable to declines in plywood prices. The gain to consumers would be mainly located in the eastern regions where the bulk of consumption takes place.

After a ban on log exports, the losers would be the stumpage producers. The bulk of the loss would be in the Pacific Northwest (west side) region and would be shared by both public and private timber owners. In addition to immediate losses, stumpage producers would have a lower incentive for intensive management practices. Reducing these practices in the near term might lead to a more constrained timber supply outlook in the long run and the attendant problems in the product markets.

In summary, the effect would be to shift income from stumpage producers in the Pacific Northwest (west side) region to processors in that region and to consumers located mostly in the East. It is difficult to make any statements about the apparent net social loss of a ban since Canadian gains and losses are not included in the computations. In addition, Japanese consumers would benefit to the extent that lower U.S. or Canadian lumber prices are reflected in the price of lumber exports.

Discussion

The effect of a ban on log exports on prices, quantities, and major social groups would not differ substantially between the question of who (Canada or the United States) becomes the main supplier of lumber to Japan. Product prices would increase in the near term because capacity expansion would lag behind the need for additional capacity. By 1985, however, capacity would expand, slowing the increase in prices for products. Capacity expansion would be fueled by increased profits resulting from lower production costs. Because production costs would continue to decline (relative to the base scenario), product prices would fall by 1990 to below the base scenario.

A ban on log exports would affect forest inventories on private lands, especially in the Pacific Northwest (west side) region, and would lower harvests in the near term, diminishing the drain on inventories and reducing the incentive for intensive forest management. By 1990, inventories would be 12 percent higher per acre for forest industry and 2 percent higher for other private owners in the Pacific Northwest (west side) region.

The two scenarios discussed in this paper should be considered as only two possible interpretations of the sequence of events likely to follow a ban on log exports. There are several reasons. The sale of an additional 4.42 billion board-feet of lumber to Japan by either U.S. or Canadian producers would require a major shift in marketing strategies for North American producers--from North American to Japanese markets. The ability of North American producers to rapidly shift market strategies is open to question. There are also questions about the ability of Japan to switch from the import of logs to the import of lumber. Major shifts in distribution and marketing channels would have to take place in Japan. There are also political questions about the possibility of a ban on exports from a Japanese perspective. For example, would Japan somehow retaliate in response to a ban on log exports?

There are questions about the ability of North American producers to respond to a ban on exports by increasing production as depicted in our analysis. For example, would capital be available and would logs be available in the right places within the Pacific Northwest (west side) region?

Because of all the uncertainties, it is possible to come up with a number of scenarios in addition to the ones discussed in this paper; but there is probably not one "most likely" scenario that everyone involved in the issue would agree with. Regardless of the sequence of events assumed to follow a ban on log exports, however, the pattern of the dis-

tribution of the impacts of the ban on stumpage owners, timber processors, and consumers seems straightforward: processors and consumers would
gain at the expense of stumpage owners, primarily in the Pacific Northwest
(west side) region. Although the pattern is clear, the size of the impacts
on any one of these groups is uncertain.

The regional nature of the impacts of a ban on log exports is another
factor that is not influenced by the assumed sequence of events following
the ban. The question of a ban on log exports is mainly regional, not
national. As such, it is intertwined in the broader issue of timber supply
in the Pacific Northwest. For example, departures from nondeclining, even-
flow harvest-scheduling policies on federal lands and a ban on log exports
are two possible policy options that could be used to offset the expected
decline in timber harvest on industry lands in the Pacific Northwest.

REFERENCES

Adams, D. M., and R. W. Haynes. In press. "The 1980 Timber Assessment Market Model: Structure, Projections and Policy Simulations," Forest Science.

Clawson, M. 1975. Forests for Whom and for What? (Baltimore, Md., Johns Hopkins University Press for Resources for the Future).

Darr, D. R., R. W. Haynes, and D. M. Adams. n.d. "The Impact of the Export and Import of Raw Logs upon Domestic Timber Supplies and Prices." Unpublished report on file at the Pacific Northwest Forest and Range Experiment Station, Portland, Oregon.

Haynes, R. W. 1976. Price Impacts of Log Export Restrictions Under Alternative Assumptions. USDA Forest Service Research Paper PNW-212, (Portland, Oregon, Pacific Northwest Forest and Range Experiment Station).

McKillop, W. J. 1974. Economic Impacts of an Intensified Timber Management Program. USDA Forest Service Research Paper WO-23 (Washington, D.C., U.S. Forest Service).

Sedjo, R. A., and A. C. Wiseman. 1979. "Log Export Prohibition: A Welfare Approach," preliminary draft. (Washington, D.C., Resources for the Future).

Stanford Research Institute. 1974. "Benefits and Costs of Alternative Log Export Policies--Phase One Report and Alternative Log Export Policies for the Long Term--Phase Two Report." Unpublished report prepared for the Pacific Northwest Regional Commission, Vancouver, Washington.

Weiner, A. A. 1973. "Export of Forest Products: Would Cutting Off Log Exports Lower Prices of Wood Products?" Journal of Forestry, vol. 71 no. 4, pp. 215-216.

Discussion by Bruce Lippke

Another Step in Understanding Trade Impacts

The Haynes-Darr-Adams paper probably represents the most sophisticated
and of econometrics ever used to unravel the probable impacts of a ban on
g exports.[1] For their attempts they deserve strong encouragement. It
ould not be forgotten that the model was designed to characterize domes-
c supply and demand markets, not exports. As a consequence the suita-
lity of extending the model to stimulate trade issues deserves thought-
l criticism from those of us who have been watching the economics of
ade at work for many years, and I do not intend to hold back on my res-
vations. There is no doubt that the simulations from their model do
udely capture some of the main economic factors at work, but I will save
ch comments until last in order to end on a positive note. I will con-
ntrate mainly on those structural limitations of the model important to
alysis of exports, the resulting model simulation problems, and the im-
ct on timber growing from the loss of wealth to timber growers.

---

[1]A yet unpublished thesis by Mich Ueda does econometrically model a
ll supply and demand analysis for three competing international regions.
is would be econometrically more complete for purposes of evaluating
ade issues.

The Market Model Structural Limitations

Since the paper does not describe the technology behind the Forest Service Assessment Market model, I will limit my comments to several aspects of the model that are particularly important for evaluating structural changes involving exports. The model is obviously much better for simulating domestic policy issues for which its coverage is more complete

Now there are a lot of elasticities in the model on both the supply side and the demand side that are subject to considerable error because of both limitations in the data and statistical estimation problems. But at least the model does try to respond dynamically, using quantified profit margins to generate expansions in capacity. Unfortunately, the model dyna mics on the timber pricing and harvest are probably the most doubtful and are inherently biased to short-term cyclic effects. The impacts of profi ability on timber and the subsequent investment in intensive management a not considered, hence some important long-term impacts are missing. More important, the model lacks the detail necessary to characterize the impac of export markets since most of the volume comes very a very small part o the total geography, namely, the port communities of Puget Sound, the Col mbia River, and Grays Harbor in Washington, with a little help from the coast of Oregon. Hemlock logs from these export regions are simply not substitutable for the plywood logs of the Willamette basin in Oregon, a substitution implied by the model.

The elasticity estimation problems, although challenging technically should not be too critical to the conclusions even if they bias the magni tude of the impact significantly.

But, the inability to model the product substitution issues can produce errors in conclusions as it applies to specific trade products. As
a consequence, the model results do need some critical examination.

But the most limiting problems are the models treatment of the Japanese market, the Canadian supply region and the alternative suppliers to the
Japanese market. The total Canadian supply is treated as an elementary
excess supply function without limitation, the Japanese market is nonexistent, and the alternative suppliers are also nonexistent. This not only
greatly limits the kinds of simulations that one might try to conduct with
the model but also raises serious questions about the implied substituta-
bility of the products being traded.

As a consequence of these limitations, the authors very properly limit
their examination to some very simple simulations. But to suggest that the
full range of impacts of a ban can be adequately examined by only two scenarios, one in which Japan replaces their log purchases by lumber from the
United States or from Canada, is far too limiting. In either case, Japan
would be forced to close down 30 percent of its lumber mills. Just imagine
what would happen in the United States if we shut down 30 percent of our
mills. In an earlier paper, the authors showed what would happen if
neither the United States nor Canada captured a major portion of the market: revenues to the Northwest's timber producers would be totally devas-
tated. Clearly a ban on more than 30 percent of Japan's solid wood needs
would generate every effort by Japan for alternative sourcing, making a
scenario with supply from other than North America no more extreme than
the two cases illustrated.

What We Should Know Versus What the Model Tells Us

The model shows that banning exports reduces the demand on Pacific Northwest (west side) timber enough to bring down stumpage prices dramatically. Obviously, product prices rise, expansion in other regions increases with higher margins, including conversion of some of the logs to lumber and plywood in the Northwest. But, strangely, the timber values in other regions do not go up much. This implies a very high efficiency in transforming existing timber inventory and the banned log exports into lumber so that the desired demand for timber in other regions increases very little.

The result is unreasonable. Hemlock logs are not directly substitutable for plywood logs, and the U.S. mill yields are typically 30 percent lower than Japanese mill yields. Hence, the derived demand for world timber to service the world's consumers will be much higher with a ban on exports. Both higher product prices and stumpage prices, especially outside of the Pacific Northwest should have been expected to source the additional timber demand.

One interesting measure of the efficiency problem is the implied reduction in roundwood harvest for chips in the Pacific Northwest. The model implies a reduced demand for roundwood of 80 million cubic feet in 1980 and over 200 million by 1990, yet the Northwest pulp industry exists almost totally on residual chips (91 percent) not roundwood (9 percent). These projections would produce a reduction in roundwood pulpwood consumption of six to eighteen times the volume currently harvested. This would appear to be a totally unfeasible solution. The problem however exists in the baseline solution, which assumes a dramatic decline in lumber production in the West with the pulp facilities sourced by an increasing harvest of roundwood. Now there is just no way that the western pulp facilities can remain competitive in the world markets by harvesting roundwood. Nor is it correct to derive wood product margins from average stumpage values as the harvest mix shifts dramatically toward a pulpwood harvest.

So the reality is that the efficiency of product substitution is nothing like that captured in the model and hence the derived demand for solid wood timber would be much higher with stumpage prices higher and above the baseline case in all but the Pacific Northwest (west side). Product prices would also be higher, reflecting these cost increases. It is inconceivable that we could get lower product prices in the United States at the same time that the Japanese demand is met by higher-cost U.S. or Canadian lumber production. But the simulation produced first higher prices and then, strangely, produced a volume response that pushed the product prices down quickly.

As another example of these problems, the paper suggests that Pacific Northwest (west side) plywood mills would not have to close down with a ban on exports. Now the plywood mills that are noncompetitive with southern plywood and the expansion in flakeboard are principally in the Willamette Basin, not an export region. An export ban will not result in hemlock logs being hauled over the mountains to source these facilities. In fact, the export mix is not a major source of plywood logs anywhere. The model simply cannot segment the market to the degree required to make that kind of judgment.

Distribution of Impacts

In one sense, the model overstates the consumer benefit since prices would be higher and consumption lower. But since I would prefer to think that the wealth transfer impact on the consumer would be consistent with the wealth loss of the timber grower, the change in price times quantity is more relevant than the price times the change in quantity. The general character of a massive wealth loss to timber growers and a transfer of part of this to producers and consumers with some loss in efficiency is correct. It would have been enlightening if it had been pointed out that the loss in welfare to the Pacific Northwest timber growers, which is much greater than the producers' benefits, is, in fact, transferred to consumers in other regions. Aside from the efficiency loss from which everyone loses, the ban produces a partially offsetting subsidy to eastern consumers, as the producer must pass his gain on to the eastern consumer in order to compete with southern and Canadian producers in markets where he is now a very poor competitor. The ban not only produces a massive wealth transfer from timber growers to producers and consumers, but also a wealth transfer from the resource region to the consumer region. Now that is absolutely contrary to the goal of energy independence. We need to encourage solar energy conversion, such as the management of forest acres, not discourage it.

Timber Impact from the Wealth Redistribution

The paper never mentions any problems with this massive wealth redistribution away from timber growers. Thirty percent of the timber owners' revenue sources would be lost. This is an enormous confiscation of returns to managing timber and an enormous loss of cash available for reinvestment.

ignore the negatives that this would have on forest management in the
st, and hence on available timber, is an omission that must be rectified.

With present timber values, most ownerships (excluding the Forest Ser-
ce) are probably intensively managing their high-site lands. Some owner-
ips have begun to do some intensive management of medium-site lands, but
ere is relatively little management of low-site lands. Now that means
at, as western markets are restricted, the 30 percent decline in values
uld shift us back toward less management in the West, while it would in-
ease the intensity of management for the South. The result over the long
n, of course, is more wood from the South and even less wood from the
st.

In the current and projected range of prices the long-term supply
sponses to prices should be very high.

A further increase in prices will provide attractive returns for in-
nsive management in the medium-site classes and some additional manage-
nt of low-site classes. Decreases in prices should produce a rollback
the intensive management of medium-site land and some decrease in the
nagement of high-site lands. The volume of growth response to price in
is range is well over 1.0 for the most affected site class since the
nagement response can produce three times the volume from natural
ands. This long-run elasticity probably averages near 1.0 across all
asses (appendix A).

Now to me, a log export ban that produces a price decline of 30 per-
nt would probably produce an equal decline in volume growth because of
duced management intensity. If the long-term growth from management was
duced by 30 percent, there would probably also be some reductions in

harvest over the shorter term since the current harvest could then not be sustained.[2] Investing more in converting facilities and infrastructure does not make a lot of sense even for the nearer term if the opportunities to grow more wood for the long term have become inferior to that available in other regions.

Conclusion

The market model simulations of an export ban do illustrate several points. The ban forces an enormous wealth transfer from the timber grower to the producer who must pass on much of that gain in order to compete with the other supply regions. There is a net loss in overall wealth. There also a wealth transfer away from the Pacific Northwest resource base to the rest of the nation.

The model is not adequate to characterize the substitution issues for the products being traded or the dynamic responses in the Japanese market

As a consequence, it clearly understates the costs of substitutions and, therefore, overstates the benefit to both producers and consumers.

And lastly, the model fails to consider the negative impact of the wealth transfer itself. With grossly lower returns, one must expect less management intensity and less growth. This should induce a reduced harvest even in the near term but certainly a major reduction in the wood available over the longer term.

As such, the model fails to characterize the economic loss of under-utilizing one of our major solar energy resources, namely western forest lands.

---

[2] The model assumes harvest cannot be sustained even in the base case but the growth assumptions in the base case are believed to be too low for current forest management levels.

Appendix A: Long-term Timber Supply

Econometric models often derive very low timber supply responses to price changes. It should be remembered that whatever the estimate, this is a short-term elasticity characterizing short-term cyclical changes. It would be a major mistake to assume that price has no impact on long-term timber management. To put this longer-term impact in better perspective, I would like to review the investment decision points that make up a supply curve in order to clarify why prices should have a major impact on supply response.

The volumes shown in figure 1 are cubic growth measures per acre per year and therefore are representative of sustainable harvest levels as a function of investment in forest management. Below some price Po, the returns for the capital tp pay for taxes and other administrative costs are below a target rate of return (such as the 8 percent real DCFROI realized by manufacturers over recent decades). As a consequence, the present value of the land, PVL, is less than zero, and private ownership for timber pur- poses might be abandoned. Public ownership would probably continue to pro- vide a conservative estimate of sustainable volume for harvest to maximize their gain (minimum loss below target return). Private owners holding the land for other speculative uses would similarly try to minimize their los- ses. So the supply curve stays at something close to Vo until prices rise to Po. Volume would not increase as prices rise slightly above Po unless there were conversions from public to less conservative private management philosophies.

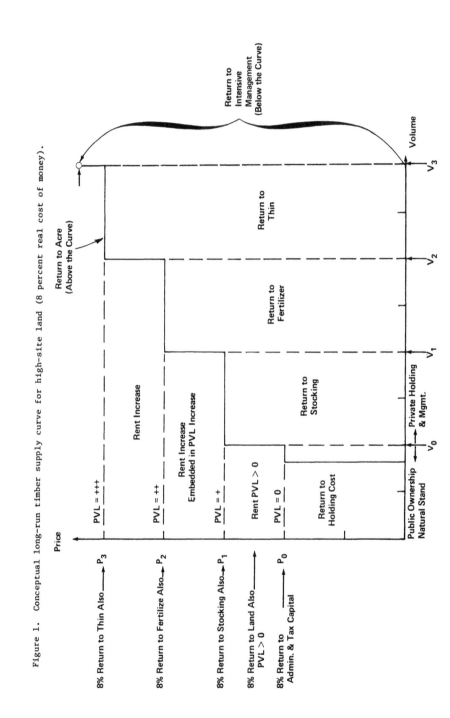

Figure 1. Conceptual long-run timber supply curve for high-site land (8 percent real cost of money).

At Po, the returns to the capital cost of administration and tax begin to exceed target returns; hence the present value of land (PVL) begins to rise above zero from Po to P1. The area PoVo is a return to the administrative cost of holding natural stands. It is neither a return to management intensification nor to rent for the land. The area (P1-Po) Vo is a rent return to the land as PVL rises and, for simplicity sake, must include the increased capitalized cost to pay for land taxes as they probably increase.

At P1, the 8 percent target return is also reached for a first management intensification step, such as stocking. The area PL (V1-VO) is the return to the capital required to invest in stocking.

From P1 to P2 the present value of the land increases further with the return to stocking already justified until returns to another management intensification step reach the 8 percent target level. This continues until the returns are exceeded for all known levels of management.

All of the revenues under the supply curve are returns to intensive management. They are the capitalized costs of site prepration, planting, and so forth. All of the revenues above the supply curve are returns to the land resource, including land, and taxes plus the returns to the land-holding costs such as insurance. (The prices shown should be considered net of yield taxes).

The Impact of a Market Restriction

Now it should be clear that if a log export market ban dropped timber prices by 30 percent within the range of Po to P3, there would be a sizable impact on forest management.

Figure 2 shows such a timber supply curve, based on the Washington State DNR Productivity Study, using a 7 percent interest on cost of money. The supply volume triples for a tripling of prices. The supply response to price is even greater than 1.0 in the midrange of prices. And the current range of prices is generally above the minimum price for positive PVL except for very low site lands.

Note that a price decline of 30 percent near the midpart of the curve would produce more than 30 percent decline in volume. Furthermore, this means that more than half of this revenue is a loss to the costs of forestry operations, the smaller half being a decline in return to the landholder and land taxes.

Another example of this supply curve was demonstrated by Clawson (1974 in which his supply curve across all owners provides a supply response to price of about 1.0 (figure 3).

These then are the issues of wealth transfers affecting timber management that are implied in the papers on the log export ban. A wealth transfer away from timber growing reduces forest management, land values, and land taxes.

It seems to me the really major impact of a market price adjustment of a market ban is a reduction in the timber volume that will be made available. While we might think of this volume reduction as being only in the long term, there is little doubt that many managers will begin to adjust their investment plans for lower rates of investment in facilities consistent with lower-growth plans.

Figure 2. Timber supply response.

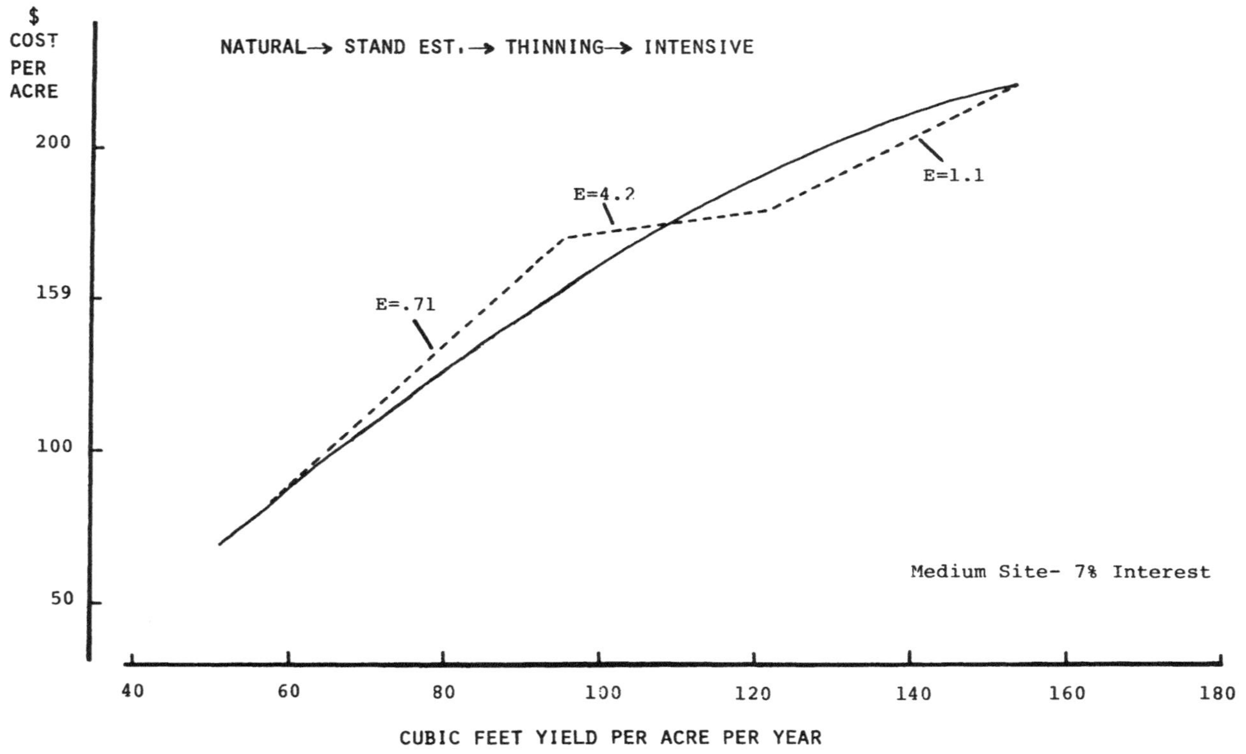

NATURAL→ STAND EST.→ THINNING→ INTENSIVE

E=1.1

E=4.?

E=.71

$

COST
PER
ACRE

200

159

100

50

40   60   80   100   120   140   160   180

CUBIC FEET YIELD PER ACRE PER YEAR

Medium Site- 7% Interest

Figure 3.   Western Washington and Oregon annual sumpage supply--all
             owners.   From Marion Clawson, <u>Forest Policy for the Future</u>:
             <u>Conflict, Compromise, Consensus</u> (Baltimore, Md., Johns Hopkins
             University Press for Resources for the Future, 1974).

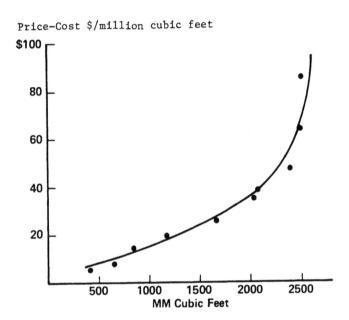

Price-Cost $/million cubic feet

And this is especially true when you think of the relatively better
opportunity in other regions where exports are a less critical part of
the markets for products derived from the forest acre.

     The Haynes-Darr-Adams paper measure the timber growers' wealth trans-
fer as the change in prices times average volume. As the intensive manage
ment curve slows, this ignores the greater half of the revenue involved,
namely the return to intensive forest management (that is, P x ΔQ). These
are not variable operating costs since it takes a half-century for pay
back; instead, they are all margins, a return to capital.

From the perspective of a region like the state of Washington, a log export ban loses both the return to land and taxes (a "producers surplus"), and the return to intensive management (costs of forestry management).

The wealth transfer may have subsidized some conversion operations for a few years at a major long-term sacrifice in growing timber which is the only comparative advantage the region has. But in order to be complete, this analysis must consider the long-term volume available and the restraint that this places on future conversion to products.

REFERENCE

Clawson, Marion. 1974. Forest Policy for the Future: Conflict, Compromise, Consensus (Baltimore, Md., Johns Hopkins University Press for Resources for the Future).

LOG EXPORT RESTRICTIONS: CAUSES AND CONSEQUENCES

Barney Dowdle

Background

Restrictions on the export of logs from publicly owned forests in the United States are in direct conflict with overall U.S. trade policy. They have been adopted for the ostensible purpose of maintaining, or creating, jobs in wood-processing industries.

Much of the political impetus for log export bans has come from the belief that the Pacific Northwest is faced with a timber "shortage" and that logs should be processed locally rather than be exported in order to save jobs. This is an exceedingly expensive myth. The belief that the Pacific Northwest has a timber shortage has its origins in the sequence of harvesting of private and public "virgin" timber, and in public policy which has not taken this economically justifiable timber-harvesting sequence properly into account. Second, transitory shifts in timber demands from the private to the public sector have encountered nonprice-responsive public timber-marketing practices and this has caused prices to escalate. Complaints about timber shortages are more complaints about these high timber prices than they are about physical shortages of wood.

The effects of log export restrictions appear to have created a market imperfection which has benefited producers in Washington State, especially the Washington State Department of Natural Resources, the Weyerhaueser Company, and a number of other producers of relatively smaller volumes of export logs.

Evidence of market imperfections is to be found in price differentials between export and domestic logs. Over the past few years the price of export logs has averaged about 50 percent more than domestic logs. Other possible reasons for price differentials are non homogeneous grade classifications and greater value-added in export than in domestic logs.

The Douglas fir region of western Oregon and western Washington, which is most affected by log export bans, has a softwood sawtimber inventory of 24 million acres which is 200 billion board feet larger than the South has on its 188 million acres of commercial timberlands. All this 200 million board feet, and probably much more, is surplus from a perspective of sound economics. The Douglas fir region could sustain an annual harvest greater than it currently produces if these excessive, over-mature inventories were harvested, and young growing stands were established in their place. The alleged timber shortage in the Douglas fir region is an artificial shortage caused by faulty public policy rather than by a physical shortage of timber.

Analyzing the effects of embargoes on log exports in the West is complicated by the fact that much of the timberland in the area is in public ownership. Publicly owned timberlands attract an inordinate amount of attention from environmentalists who would like to put them into parks and wilderness areas; many decisions affecting public timberlands are made by elected officials whose primary interests are in other industries in other areas; and the theory of public forest management does not provide estimates of costs and benefits of various management alternatives which permit them to be properly weighed, either against each other or against nonforestry alternatives. This situation has created a great deal of public misunderstanding. It has also created an environment in which there is a tendency to resort to policy experimentation. This situation has resulted in a proliferation of restrictions. More restrictions are added to compensate for those which have failed in the past.

The Log Export Problem

The growth in trade between the United States and Japan, which has characterized the post-World War II period, has significantly affected the forest products industry in the West during the past several years. Log export volumes from Washington and Oregon were less than 1 billion board-feet in 1963. By 1978 export log sales were greater than 3 billion board-feet. This represented 34 percent of the timber harvest in Washington State, and about 12 percent in Oregon (table 1).

Log exports from Washington exceeded those of other states because much of the commercial forest land in western states is in public ownership and with the exception of state-owned land in Washington, log exports are banned from publicly owned timber. Federal timber cannot be exported, nor can timber from state lands in Alaska, California, Idaho, and Oregon. British Columbia also restricts the export of logs.

In addition to log export restrictions, regulations also exist which preclude domestic producers from exporting their own (fee) timber and sub-

Table 1. Timber Production and Log Exports from Washington and Oregon,
1963-78

(MMBF scribner)

	Washington			Oregon		
Year	Total production	Export sales[a]	Percentage export	Total production	Export sales	Percentage export
1963	5,428	575	11	8,676	297	3
1964	6,241	555	9	9,418	371	4
1965	6,522	706	11	9,394	283	3
1966	6,075	809	13	8,921	389	4
1967	5,936	1,161	20	8,357	502	6
1968	6,971	1,475	21	9,743	653	7
1969	7,004	1,414	20	9,150	561	6
1970	6,495	1,697	26	7,980	637	8
1971	6,450	1,320	20	9,028	517	6
1972	7,081	1,913	27	9,630	730	8
1973	7,809	1,915	25	9,366	806	9
1974	6,876	1,618	24	8,361	766	9
1975	6,185	1,595	26	7,371	798	11
1976	6,968	1,974	28	8,153	944	12
1977	6,592	2,025	31	7,525	881	12
1978	6,983	2,369	34		931	

[a]Export sales estimates provided by Washington State Department of
Natural Resources. All other data from Florence K. Ruderman. 1979.
Production, Prices, Employment, and Trade in Northwest Forest Industries
Portland, Oregon, U.S. Forest Service, Pacific Northwest Forest and
Range Experiment Station).

stituting public timber in their mills. Much of the federal legislation banning log exports reflects the efforts of the late Senator Wayne Morse of Oregon. More recently, Representative Don Bonker of Washington has been the champion of log export bans.

The effect of the Bonker-Morse constraints has been to shift most of the demand for export logs to Washington, and, perhaps more important, to segment the log market. The export and domestic log markets are suffi ciently insulated from each other to permit multiple-pricing. Export log prices tend to be higher than domestic log prices, sometimes by over 100 percent (table 2).

The price differential shown in table 2 may reflect differences in value added--export logs may have gone through more processing than domes tic logs--and nonhomogeneous grade categories which permit further grade breakdowns within categories. The most plausible explanation, however, would seem to be that the Bonker-Morse contraints have created market im-perfections which lend themselves to multiple-pricing policies. Models o multipricing are to be found in most intermediate level texts in price theory, and they can be used to provide insights into the log export question.

A Static Analysis of the Log Export Market

Differentials between log prices in the export and domestic markets can be explained by the static analysis diagramed in figure 1. This ana-lysis, it should be emphasized, is merely designed to be illustrative of what the Bonker-Morse constraints appear to have achieved; a market im-perfection which lends itself to exploitation.

Table 2. Log Prices for Domestic and Export Sales in Western Washington
and Northwestern Oregon, 1963-78

($/million board-feet scribner)

| | Species | | | | | |
| | Douglas fir (all grades) | | | Western hemlock (all grades) | | |
Year	Domestic sales	Export sales	Percentage difference	Domestic sales	Export sales	Percentage difference
1963	57.90	61.60	6	46.90	56.80	21
1964	57.70	68.50	19	47.60	62.20	31
1965	61.50	72.50	18	49.30	70.00	42
1966	62.30	77.10	24	39.80	76.00	53
1967	61.80	80.00	29	49.20	80.60	64
1968	75.00	93.20	24	62.60	96.00	54
1969	88.10	122.40	39	78.40	122.40	56
1970	73.50	116.90	59	63.30	120.80	91
1971	86.20	111.50	29	72.50	111.70	54
1972	86.80	125.40	44	85.50	129.10	51
1973	137.30	284.00	107	112.90	286.00	153
1974	159.10	247.20	55	142.10	240.30	59
1975	154.70	217.40	40	134.00	206.10	54
1976	180.40	250.40	39	145.90	241.40	65
1977	227.20	265.30	18	172.30	258.40	50
1978	250.00	321.60	29	182.90	282.60	55

Source: Florence K. Ruderman, Production, Prices, Employment, and
Trade in Northwest Forest Industries (Portland, Oregon, U.S. Forest Service, Pacific Northwest Forest and Range Experiment Station, 1979).

Let D_h and D_x be demands in the domestic (home) and export markets,
respectively. The separation of the two markets is mandated by the Bonker-
Morse constraints. Most firms in the Pacific Northwest are now at least
partially dependent upon federal timber. Because of substitution provisions in the law, those firms which purchase federal timber are effectively precluded from exporting. "Grandfather" clauses in the law and regulations permit some exporting of fee timber while purchasing federal timber,
but not in amounts sufficient to arbitrage prices between the domestic and
export markets.

On the supply side of the export market, three components of supply
are identified in figure 1: (1) fringe competitors, which include mostly
nonintegrated timber holdings, farm and woodlot owners, and fee timber
exported by vertically integrated firms under "grandfather" clauses; (2)
public agencies which regulate timber harvest on the basis of the "sustained yield" principle, for example, the Washington State Department of
Natural Resources (DNR); and (3) a dominant firm in the log export trade
which has sufficient timber holdings to supply its own mills and provide
logs to the markets, most of which are exported.

Supply for fringe competitors is labeled S_f in figure 1. If public
agency supply is added to S_f, the S_{fp} is obtained.[1] Marginal cost of output for the dominant firm is labeled MC_d.

---

[1]The DNR is to some extent price responsive, as evidenced by timber
price, and the agency's timber sales statistics. It would not be realistic, however, to construct figure 1 as if the DNR were a profit maximizer,
such as assumed in supply analysis. The DNR's annual allowable sale (harvest) relationships, which are subject to ad hoc administrative adjustments, according to market and other conditions, are far removed logically
and empirically from the supply curve of economic theory.

Figure 1.

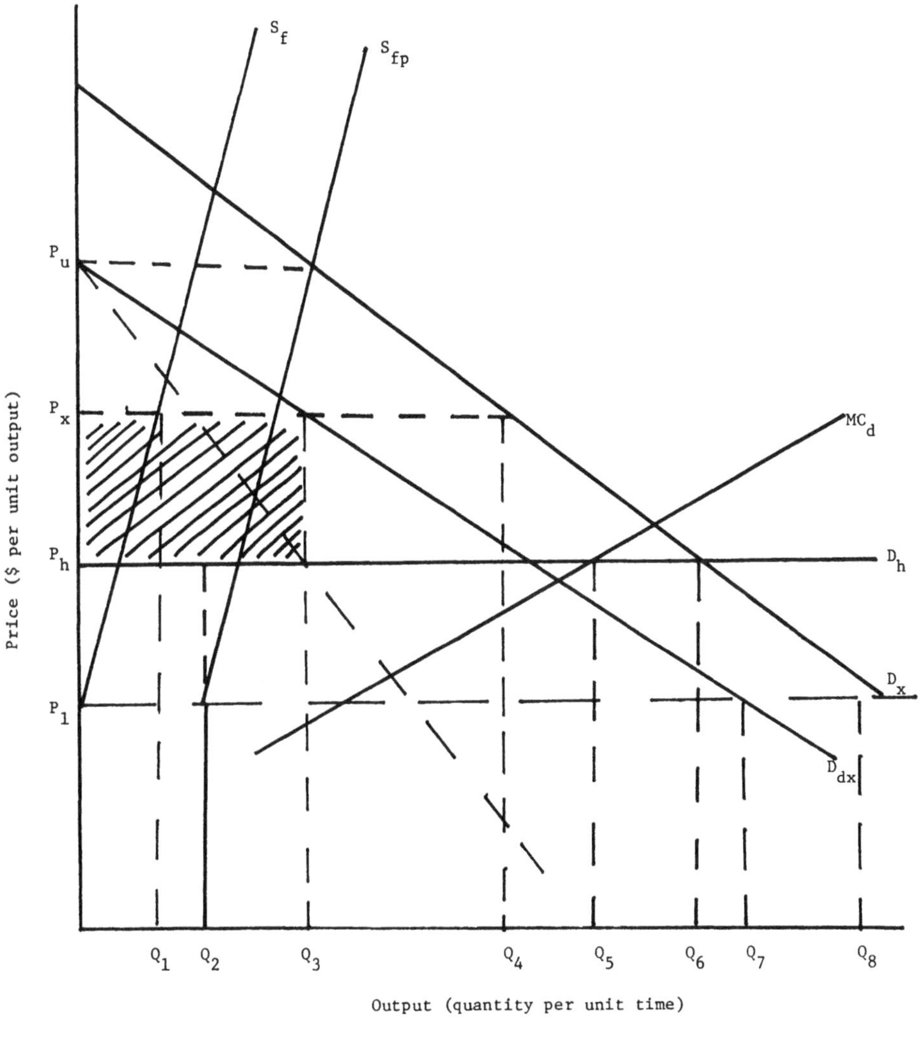

Price ($ per unit output)

P_u

P_x

P_h

P_1

S_f

S_{fp}

MC_d

D_h

D_x

D_{dx}

Q_1 Q_2 Q_3 Q_4 Q_5 Q_6 Q_7 Q_8

Output (quantity per unit time)

255

The construction of figure 1 proceeds as follows:

1. S_{fp} is subtracted from D_x to get D_{dx}, residual demand for the dominant firm in the export market. At price P_u, the dominant firm would sell nothing in the export market; fringe competitors and public output would clear the market. At price P_1, the fringe competitors would sell nothing in the export market, and the difference between D_x and D_{dx} is the output of the public sector.

2. Marginal revenue for the dominant firm in the export market (MR_{dx} is derived from D_{dx}.

3. Marginal cost for the dominant firm (MC_d) is positional to reflect the fact that the dominant firm has lower marginal costs in the export market than fringe competitors; marginal cost for the public agency is assumed not relevant to analysis.

4. Demand in the domestic (home) market is constructed to reflect the fact that the output of producers in the export market is small relative to total output in the domestic market. As a result, demand is highly elastic in the domestic market.

The interpretation of figure 1 is relatively straightforward. The dominant firm is assumed to be a profit-maximizer, hence it will equate marginal cost and marginal revenue. Moreover, marginal revenue must be equal to marginal cost in both the domestic and export markets.

Marginal cost of the dominant producer is equal to marginal revenue in the domestic market at P_h; marginal revenue being equal to price in this market because demand is assumed to be perfectly elastic. This occurs at an output level Q_5. Marginal revenue in the export market for the dominant firm is equal to P_h at Q_3 units of output.

To maximize profits the dominant firm will, therefore, produce Q_5 its of output, Q_3 of which are sold in the export market. Selling price the domestic market is P_h; in the export market it is P_x. The cross-tched area in figure 1 indicates excess profits realized by the dominant rm as a result of the separation of the domestic and export markets.

Output of the fringe competitors in the export market is Q_1, where is equal to P_x, and the public sector produces Q_2 units for the export rket. Total export sales equals Q_4 units.

What are the implications of the foregoing analysis? Prior to addres-ng this question, it should be reemphasized that the analysis is static, d that demands in both the domestic and export markets are relatively stable over time. In addition, the marginal cost for the dominant firm d the supply of the fringe competitors will shift over time as timber ventories are adjusted to long run equilibrium levels. While the private ctor in the Pacific Northwest appears to be nearing completion of its d growth ("virgin") timber inventory adjustment, this process is still derway. The most reasonable assumption to make in this regard is that th S_f and MC_d will rise (shift to the left): output per unit time will ll as old growth inventories are depleted.

The extent to which public sector output will change depends upon the tent to which current annual allowable sales (harvests) can be maintain-. The DNR claims that it is currently on a "sustainable harvest" program ich is, indeed, sustainable; hence, there is no reason to assume public tput will change.[2]

It can be noted that if the timber inventory of the public sector not currently at an equilibrium level, and if output is assumed to re-in constant over time, then profits are not being maximized. Sacrificed fits would, in this case, be a measure of the cost of following a stained yield" timber management policy.

Subject to these qualifications, the conclusions of the static analysis described above which are of interest are:

1. Fewer logs are exported because of the Bonker-Morse constraints than otherwise would be. This difference, as shown in figure 1, would be $(Q_6 - Q_4)$.

2. Fringe competitors in the export market are producing more and selling it at higher prices than they would if constraints did not exist. Output is determined by the intersection of S_f and P_x. In the absence of any constraints it would occur at the intersection of S_f and P_h.

3. If MC_d rises, the dominant firm would reduce domestic sales before it would reduce export sales.

4. Income transfers are from buyers of export logs to the sellers of export logs, that is, fringe competitors, the dominant firm, and the public agency. If the latter produced more, then log prices in the export market would be expected to fall. What this might do to the relationship between marginal cost and marginal revenue for the public agency is beyond the scope of the present analysis.

Discussion by William McKillop

Three main issues are raised by Barney Dowdle's paper.

1. What degree of relevance does National Forest timber output policy
   have for the log export issue?

2. Is the price for export logs higher than that for domestic logs
   because of the marketing strategy of the dominant private exporter?

3. What gains and losses, regionally and nationally, would be associated with a removal of the current restrictions on the export
   of logs from federal lands in the West.

National Forest Timber Output Policy

The pros and cons of altering the planned levels of National Forest
timber harvest is too large a topic to discuss in a workshop on international trade. Issues such as financial criteria for setting output goals,
nontimber benefits of the National Forests, and legislative and political
constraints would quickly dominate the entire discussion. A symposium on
timber output policy should consider, among other things, the effect on
log exports, but the reverse is not necessarily the case. Higher levels
of timber output from the National Forests would, of course, lessen the
pressure for an export ban. But in dealing with the log export question

per se, it is almost essential to take National Forest output policy as given if issues are to be analyzed in a meaningful way.

Disparity Between Export and Domestic Prices

The disparity between export and domestic log prices is indisputable, but the reason for the disparity is open to question. Economic theory suggests that a seller who is a monopolist in each of two distinct markets could maximize net returns by setting sales levels such that marginal revenues in each market are equal to each other and to overall marginal cost (Henderson and Quandt, 1971). Price will be lower in the market with the greater demand elasticity. Intuitively, it appears that the theory can be extended to the case where the seller is a price-taker in one market, in this case the market for domestically processed logs. One must, however, ask whether this normative theory is an accurate portrayal of what the major supplier of private supplies of export logs is willing and able to do in reality. In the absence of further information, the theory must be regarded only as an untested hypothesis.

Clearly, there could be other reasons for the disparity when average export price is compared with average domestic price. The product mix in the two markets is unlikely to be similar. Lower grades are less transportable from an economic point of view, and one would therefore expect domestically produced logs, on the average, to be of a lower grade. Even if export logs have a higher price grade-for-grade, it must be noted that though the logs may be physically identical, they may be different commodities from an economic point of view. Domestic prices tend to be spot prices, whereas most export log prices are established under a contractual arrangement, and may reflect a premium to ensure stability of supply for the buyer. Furthermore, export log prices have a value-added component that is not contained in domestic prices. These reasons may not explain fully the domestic/export disparity, but they call into question the view that the disparity is caused by the exercise of monopoly power by the major seller of export logs.

Gains and Losses from Removal of Export Restrictions

If the current restriction of U.S. log exports is removed, the following primary market effects will be observed: (1) there will be an increase (rightward shift) in supply of logs to the export market, coupled with a price decrease; (2) there will be a decrease in the supply of logs, lumber, and plywood by U.S. producers to domestic markets, and an associated price increase; and (3) there will be an increase in the supply of lumber by Japanese wood processors to the domestic Japanese market (and a corresponding price decrease).

These primary effects may be offset to some extent by secondary effects, such as a decrease in the supply of Canadian and U.S. lumber to Japanese markets and a shift in Canadian exports from Japan to the United States, but they will lessen, rather than eliminate, the primary effects noted above. As a result there will be:

1. A more favorable balance of trade for the United States.

2. A gain in consumer surplus by Japanese consumers.

3. A loss in consumer surplus by U.S. consumers.

4. A reduction in output of processed wood and a decrease in timber industry employment in the Pacific Northwest.

5. An increase in stumpage prices (and presumably revenues) for those timber growers such as the U.S. Forest Service who are currently prohibited from exporting logs.

6. A decrease in stumpage price and revenues for timber growers who are currently exporting logs.

Of these effects, the most acute, in terms of visibility and political debate, will be timber industry employment and the survival of less competitive wood-processing firms in the Northwest. Darr (1975) estimated that there were some 4.7 person-hours of employment per 1,000 board-feet of logs exported as opposed to 12.6 person-hours for each 1,000 feet processed into lumber in the Pacific Northwest. The corresponding figure for plywood and veneer was 19.5 person-hours. Thus a diversion of logs from the domestic to the export market could lead to a significant loss in timber industry employment. Stevens (1979) has suggested that regional losses in timber industry jobs, because of labor force mobility, are less crucial than one might believe. But mobility, undoubtedly, varies with age and occupational class, and depends on job availability in other areas, and in other sectors of the economy. Anticipated declines in timber output in western Washington and western Oregon suggest that jobs may not be available for those workers displaced by changes in log export policy.

On balance, it would appear that retention of existing log export restrictions is desirable. This does not mean that extension of restrictions is advocated. Nor does it mean that some groups would not gain from a total removal of the log export ban. But any windfall gains will tend to be diffused. Whereas windfall losses will be focused and borne by a comparatively small group in society.

263

REFERENCES

arr, David R. 1975. Softwood Log Exports and the Value and Employment Issue. Research Paper PNW 200 (Portland, Oregon, U.S. Forest Service, Pacific Northwest Forest and Range Experiment Station).

enderson, James M., and Richard E. Quandt. 1971. Microeconomic Theory, 2nd edition (New York: McGraw-Hill).

tevens, Joe B. 1979. "Six Views About a Wood Products Labor Force, Most of Which May Be Wrong," Journal of Forestry vol. 77, no. 11, pp. 717-720.

Appendix A

AGENDA

RESOURCES FOR THE FUTURE/AMERICAN FORESTRY ASSOCIATION
WORKSHOP
ISSUES IN U.S. INTERNATIONAL FOREST PRODUCTS TRADE

Date: March 6 and 7, 1980

Place: American Forestry Association, 1319 18th Street, N.W.
            Washington, D.C.

Morning Session: 9:00 a.m.	THE FUTURE ROLE IN WORLD FOREST RESOURCE TRADE Moderator: Emery N. Castle, RFF
David Darr--USFS	"U.S. Exports and Imports of Some Major Forest Products: The Next Fifty Years"
Roger Sedjo--RFF	"Exotic Forest Plantations: What are the Impli- cations for U.S. Forest Products Trade"
Panel Discussion:	John Zivnuska, University of California Dwight Hair, U.S. Forest Service Jan Laarman, North Carolina State University John Ward, National Forest Products Association

******

Afternoon Session: 1:45 p.m.	THE EFFECT OF RESTRICTIONS UPON U.S. FOREST PRODUCTS TRADE Moderator: Rex Ressler, AFA
Darius Adams/Richard Haynes--USFS	"U.S.-Canadian Lumber Trade: The Effect of Res- trictions"
Samuel Radcliffe--RFF	"U.S. Forest Products Trade and the Multilateral Trade Negotiations"
Panel Discussion:	Irene Meister, American Paper Institute Harold Wisdom, Virginia Polytechnic Institute Louis Vargha, The Weyerhaeuser Company A. Clark Wiseman, Utah State University

******

Second Day

| Morning Session: 9:00 a.m. | THE ECONOMIC EFFECTS OF LOG EXPORT RESTRICTIONS Moderator: Marion Clawson, RFF |

Morning Session: THE ECONOMIC EFFECTS OF LOG EXPORT RESTRICTIONS
  9:00 a.m.   Moderator: Marion Clawson, RFF

A. Clark Wiseman/   "Welfare Economics and the Log Export Policy
Roger Sedjo--USU-RFF   Issue"

Richard Haynes/ David
Darr/ Darius Adams--USFS   "U.S.-Japanese Log Trade: Effect of a Ban"
Oregon State University

Barney Dowdle--Univer-   "Log Export Restrictions: Causes and Conse-
sity of Washington   quences"

Panel Discussion: Philip Cartwright, University of Washington
    Bruce Lippke, Weyerhaueser Company
    William McKillop, University of California

****

Appendix B

PARTICIPANTS

RESOURCES FOR THE FUTURE/AMERICAN FORESTRY ASSOCIATION WORKSHOP
ISSUES IN U.S. INTERNATIONAL FOREST PRODUCTS TRADE
March 6 and 7, 1980

Darius Adams
Department of Forestry
Oregon State University

Keith Aird
Canadian Forestry Service

Walter C. Anderson
Southern Forest Experiment Station
U.S. Forest Service

Carl Bernsten
Society of American Foresters

Robert Buckman
Deputy Chief for Research
U.S. Forest Service

Joseph Buongiorno
Department of Forestry
University of Wisconsin

Warren Calow
Department of Industry, Trade,
and Commerce
Ottawa, Ontario

Philip Cartwright
Department of Economics
University of Washington

Emery Castle
President
Resources for the Future

Marion Clawson
Resources for the Future

Ralph Colberg
Senior Woodlands Business Analyst
The Mead Corporation

David Darr
Pacific Northwest Forest and Range
Experiment Station
U.S. Forest Service

Joseph Dose
Chief, Division of Forestry
Bureau of Land Management
Department of the Interior

Barney Dowdle
Department of Forestry
University of Washington

Robert Edwards
Manager, Division of Technical Ser-
vices
Department of Natural Resources
State of Washington

Robert Forester
Canadian Forestry Service

Edward Furlow
Wood Products Division
International Trade Commission

Alberto Geotzel
National Forest Products Association

Dwight Hair
U.S. Forest Service

Ian Hardie
Department of Agricultural Economics
University of Maryland

Richard Haynes
Pacific Northwest Forest and Range
Experiment Station
U.S. Forest Service

William Hoffmeyer
Wood Products Division
International Trade Commission

Jan Laarman
School of Forest Resources
North Carolina State University

Bruce Lippke
Manager, Marketing and Economic
  Research
The Weyerhaeuser Company

Edward Lyons
Champion International

William McKillop
Department of Forestry
University of California,
  Berkeley

James McKinney
Graduate School of Business
  Administration
Harvard University

Irene Meister
Vice President, International
American Paper Institute

John Muench
Director, Economics
National Forest Products
  Association

Richard Pardo
Programs Director
The American Forestry Association

Dean Quinney
Deputy Director
Forest Resource Economics Research
  Staff
U.S. Forest Service

Samuel Radcliffe
Resources for the Future

Rex Ressler
Executive Vice President
American Forestry Association

Clark Row
U.S. Forest Service

Roger Sedjo
Director, Forest Economics and
Policy Program
Resources for the Future

Art Smyth
Vice President, Washington Affairs
The Weyerhaeuser Company

John Spears
World Bank

James Talbot
National Academy of Sciences

Louis Vargha
Director of International Planning
The Weyerhaueser Company

John Walker
Director of Resource Services
Simpson Timber Company

John Ward
Director, International Trade
National Forest Products Associatio

Harold Wisdom
Department of Forestry
Virginia Polytechnic Institute

A. Clark Wiseman
Department of Economics
Utah State University

Robert E. Wolf
Congressional Research Service
U.S. Library of Congress

James G. Yoho
International Paper Company

John Zivnuska
Department of Forestry and Resource
  Management
University of California, Berkeley

www.ingramcontent.com/pod-product-compliance
Ingram Content Group UK Ltd.
Pitfield, Milton Keynes, MK11 3LW, UK
UKHW020858280225
455677UK00006B/96